Wild Times

WILD TIMES

TALES FROM SUBURBAN SAFARIS

BY TIM HARRISON

Ske,
THANKYOU!!

ORANGE FRAZER *PRESS*
Wilmington, Ohio

ISBN 1-882203-77-1

Additional copies of *Wild Times: Tales From Suburban Safaris* or other Orange Frazer Press books may be ordered directly from:

Orange Frazer Press, Inc.
Box 214
37 ½ West Main Street
Wilmington, Ohio 45177

Telephone 1.800.852.9332 for price and shipping information
Web Site: www.orangefrazer.com

Illustrations by Tim Harrison
Cover photo by Robert Flischel
Cover design by Tim Fauley

Library of Congress Cataloging-in-Publication Data

Harrison, Tim, 1956-
Wild times : tales from suburban safaris / by Tim Harrison
p. cm.
ISBN 1-882203-77-1
1. Wildlife rescue. 2. Pests--Control. 3. Wild animals as pets. 4. Exotic animals. I.
Title.

QL83.2 .H37 2001
639.9--dc21

2001034022

To my mom and dad who allowed me to be who I am; to my younger brother, Jim, who I will always look up to; to my sons, Adam, Alex and Aric, who I love more than life itself; and to my wife, Patricia, who has adapted to my lifestyle. I will always have a special love for you.

Acknowledgements

Special thanks to Tracy Martin who typed out my hieroglyphics; Bill Cacciolfi and Keith Gad, good friends and fellow adventurers; Bob Cranston, who opened up a whole new world to me; and to all the animal control, police, fire, rescue, and nature center employees who are there when we need them.
A very special thanks to Matt Hieb and Dorothy Lane Market.

Contents

Introduction

PEOPLE WHO LIVE IN THE SUBURBS believe that their chances of stepping on a cobra or coming face to face with an African lion are slim. The stories in this book, however, are true stories. They happened to people just like you, in neighborhoods just like yours. They will entertain you and possibly frighten you. So before reading further, grab both your snake stick and your courage and let's go on a suburban safari.

WILD TIMES

o • n • e

Snakes In The Surburban Grass

Thirty-foot Python Terrorizes Community And Other Extravagant Stories

FOR EACH OF NEARLY THIRTY YEARS, I have received an average of sixty calls about snakes in people's homes or on their property in the suburbs of Dayton, Ohio. Twenty of these calls are about exotic snakes, and five of that twenty are venomous. Most of the calls are relayed to me from local police or animal control departments. All the exotic snakes—pythons, boas, vipers, cobras—are or were someone's "pet," which means they were bought from animal dealers, pet stores, or through the local classified ads.

The snakes escaped or they were turned loose by the owner, who thought they could live in the wilds of Ohio suburbs, not understanding that our cold winters would kill them or that scared neighbors would chop them up with a hoe, or have their dogs tear them apart, or worse, some innocent child or adult would be bitten, seriously hurt or killed.

There are many urban legends about snakes, and most of them are lies or exaggerated truths. An exaggerated truth is where a three-foot snake, after many separate people tell the story to the news media, becomes a twenty-foot snake. The media then finds an expert who warns of the danger to your children, and to life on earth as we know it.

I shudder to hear the term "expert," and I tell people if someone refers to himself or herself as an expert, ask for references and a blood test; the expert may be more dangerous than the snake.

"Thirty-foot Python Terrorizes Community," read one headline, which was ominously subtitled, "Children and Pets at Risk of Being Eaten Alive." My brother, Jim Harrison, an internationally known herpetologist and owner of the Kentucky Reptile Zoo near Natural Bridge, Kentucky State Park, and I were surprised to hear an expert had told a local newspaper that a snake which had escaped from its owner—a local motorcycle gang member—was over thirty feet long and very dangerous to children and pets.

This monstrous creature would search out neighborhood children as well as small pets for food, and the community should be scared, very scared. Posses were formed to find this evil creature, and naturalists from the local nature centers were brought in. They spent a week holding hands and searching grassy fields miles from the owner's house. A panic fell over the city.

Jim and I decided to get involved. Jim went to the house from which the snake had escaped and where it was last seen. He went to the building behind that house, a motorcycle workshop. It was January, and this building had a furnace to keep the work area warm. Jim peered behind the furnace and there was the monster snake.

The experts had been searching for the snake for two weeks. Jim took thirty seconds. He pulled out the mighty beast that had the city paralyzed in fear. It was twelve feet long, docile, and in poor condition. It was a Burmese python, and since there are no laws against either stupidity or the prevention of cruelty to reptiles, the snake was given back to the owner, over our objections.

Yes, the snake was a python.

Yes, it was a large python.

And, yes, it escaped.

But that was the end of the truth.

No, snakes do not eat people.

No, it was not thirty feet long.

And, no, it was not a danger to the city.

But why merely the truth when you can plunge off into the exciting world of media extravagance and hyperbole? If the experts in the news media had even opened a book on snakes, they would have learned that there is no record that a thirty-foot-long snake has ever existed. The New York Zoological Society has had in existence for seventy-five years a "bounty" that the institution offers to anyone bringing forth proof that a thirty-foot long snake exists. There is, of course, speculation that a snake in excess of thirty feet does exist, but none of the sightings have been documented.

I have learned that when someone sees a snake or any unusual exotic animal they have the "fisherman mentality," which asserts that anything approximately two feet long automatically becomes ten feet long and dangerous. By the time the story makes it to the local tavern, the animal has acquired a record length, is deadly to civilization itself, and the storyteller was extremely fortunate to escape with his life.

King Visits Garage, Pinned By Exuberant Fans

ON THE OTHER END OF THE SPECTRUM, there are situations where you expect exaggeration and find the truth more dangerous and bizarre than fiction could ever be.

My brother and I received a call from a local police department who reported that a king cobra was loose in a garage in a home in their heavily populated downtown area. As my brother and I were both officers responding in a cruiser, we heard on the radio that both zoos in the communities near Dayton—and all the nature centers and local museums—would not respond to a call about a loose, possibly venomous, snake.

When we approached the residence, there were up to 400 people surrounding the garage and at least four cruisers from the local departments. Jim and I entered the garage with a local television news photographer following behind. Jim found the cobra behind a folded card table, pinned it down, and grabbed the snake just below the head. The capture took most of twenty seconds.

The snake was a monocle cobra and its venom was neurotoxic and there was no first aid or antivenin for this creature's bite in the tri-state area. The snake was close to five feet long and how it got into the garage is still a mystery.

The owner of the garage came home from shopping, stepped out of her vehicle, and felt something move under her foot. She immediately saw it was a snake and jumped clear as the snake puffed up and spread its hood.

She ran into the house and told her husband, who laughed and grabbed a broom, thinking it was only a garter snake. When he went into the garage he was surprised to discover a five-foot, fully puffed up and spread-out cobra, acting very aggressively.

He ran back into the house and called his local police department. That call finally ended up with Jim and me becoming local heroes for saving the city from this deadly king cobra, and the news photographer received second place in the nation in the spotlight news category for his excellent filming of the capture.

As a side note, the neighbors told us that numerous cats and a few dogs had been found dead in that neighborhood over the past few weeks. It was luck that no children had come upon the cobra, or there would have been a sad ending to an otherwise exciting story.

Large Mammal Menaces Authorities

THE CITY WHERE I WORK as a police officer, firefighter, and EMT-paramedic, Oakwood, is a southern suburb of Dayton, Ohio, which is a large city with the large city problems such as drugs and drug dealers. As a police officer, I have had the unique opportunity to be called out to assist a local SWAT team on a most unusual drug house raid—one that involved exotic snakes.

The drug raid was on the east side of the city at a well-known crack house, and the team needed a snake-handler because its inside source said there were poisonous snakes on the property and the owner would use them to attack the police if they tried to enter.

I was told to look for the police cruiser when I pulled up in my personal vehicle. I drove around the block twice and did not see anything that looked like a police vehicle. As I drove by a third time, a large, well-muscled man wearing a bandana on his head leaned out of a beat-up van and said, "Harrison."

I slowed down, backed up, and discovered that it was the SWAT members, undercover, in the van. The man with the bandana told me the scenario, and the raid began with the team announcing its presence, kicking in the front door, and throwing in percussion bombs.

People were running and jumping out of windows. A large woman wearing nothing but panties came running out the door, and one of the SWAT team members said, "Go get her, Harrison!"

"I'm here for the reptiles, not a large mammal," I said.

As the team entered, the female owner began to throw snakes at the team yelling, "They're poisonous!"

I caught one and found the snakes were pythons, Royal and Burmese. The house was secured and the bad guys arrested. I talked to the team about the snakes and what to look out for on future raids—vipers or scorpions in moneybags or stashed in drug containers.

One of the team members told me about a suspect that was supposed to have two king cobras guarding his stash of drugs and money. He said an undercover officer had seen them. I found out the name of the business and the address that the suspect used as a front. I had a friend who looked the part of a motorcycle gang member and asked him to go to this business and ask about the cobras.

When my friend, who is also a herpetologist, gained the suspect's trust, he saw the cobras.

"Tim," he said, "they are king cobras and this guy is extremely dangerous."

We are all awaiting the call from the SWAT team to bust this guy and every night saying a prayer for the team members if we have to deal with this psycho and his menagerie.

Cleverly Disguised Beast Shocks Unsuspecting Herpetologist

ONCE, I WAS CALLED TO a town north of Dayton where a forty-foot python was reported to be under a house in a well-populated suburban area. The local police and animal control officer, as well as naturalists from a local nature center, said, "This animal is a danger to the citizens." No one wanted to try and capture this monster.

Upon my arrival I was greeted by the above-mentioned group, as well as a large crowd of spectators who were very enthusiastic at the idea of someone crazy enough to crawl under the house and capture a beast of mythical proportions.

I grabbed a flashlight, looked under the house, and immediately the beam from the light shined on what appeared to be the largest python I have ever seen or heard of. I stood up and contemplated my next move as the crowd started to get more excited.

I asked one of the firemen for a pike pole, a long pole that resembles a whaling harpoon with a point and a hook on the end, and I crawled cautiously under the house. As I approached the python, I observed that the pattern of the snake was like nothing I had ever seen before. I reached tentatively forward with the pike pole and hooked the creature. It did not move, but the hide seemed to pull off the animal.

I pulled the hide back towards myself and found that it was a camouflage jacket that had been wrapped around some old tires. When I crawled closer I discovered some camouflage pants around another tire, which made the tires resemble a mammoth python, curled and ready to strike.

I decided to give the crowd a thrill. I wrapped one tire back up with the jacket and crawled out, pulling it with the pike pole. As I exited the space under the house, I yelled, "Look out!" pulling the tire violently out from under the house and causing the police, firefighters, naturalists, animal control personnel, and spectators to scream and jump completely out of their skins.

When they saw me laughing and noticed what the python really was, everyone began to laugh and the tension was broken as everyone headed back to their homes and jobs with a good story to tell their friends and relatives about the monster camouflaged snake under the house.

Creatures Evacuate Apartment Complex

HUMANITY HOLDS AN amazing fear of snakes. Once, I was dispatched to an apartment building on a report of a massive, possibly venomous snake loose in the building. When I arrived, a crowd of people were standing outside the apartment building, huddled close to each other.

As I pulled up, the crowd rushed over to my cruiser and everyone started talking at the same time, each relating his or her close encounter with this deadly creature. Finally, I learned in which apartment the snake was last seen, and I entered the building with my trusty pillowcase and snake stick in hand.

When I entered the apartment, I saw a large pot that had either been dropped or placed on something on the carpeted floor. The pot was large but if a snake was under this pot it could be no larger then three feet long, and the witnesses said it was "mammoth" in size.

I slowly lifted the pot and the snake shot out from under it and streaked across the floor. I pinned the snake down and picked up the creature that had evacuated a four-unit apartment complex.

The snake was a garter snake, which is a local snake that is nonvenomous. I showed all the people standing outside, then removed the snake from the property, and released it in a nearby nature reserve.

The snake was close to three feet long and was more scared of the people and the situation than the apartment dwellers themselves.

Once, I was called to the local children's hospital about two boys that may have been bitten by venomous snakes. The boys had been transported to the emergency room by helicopter, and the doctors were having trouble identifying the snakes by reference books.

I always teach in my seminars to schools, medical personnel, and nature centers that looking at a picture in a book may not be the right thing to do. Pictures in books are of snakes or animals that are usually captive grown and in perfect condition. Snakes in the wild look wild, and they may come in different colorations than those noted in a manual.

A man in Florida was bitten by a coral snake that was almost black in color, which was a deviation or mutation, not the typical "red to yellow kill a fellow, red to black venom lack" taught by the manuals. From the color, the man thought it was a harmless water snake and misidentified the true snake.

Add that people have exotic venomous snakes as pets, which look and act like some of our local nonvenomous snakes, and one has an even more dangerous situation, causing alleged experts and medical personnel to misdiagnose a true venomous snake bite.

When I arrived at the emergency room, I was greeted by the emergency room nurses and doctors who immediately brought the container with the snakes in it. I opened the top and let out a deep breath of relief. The snakes were northern water snakes, local snakes that are nonvenomous but extremely aggressive and act like water moccasins that are venomous but do not live in Ohio, unless someone has them as pets.

Everyone was relieved and I was scheduled to do a seminar for the emergency room personnel about venomous and nonvenomous animal bites. The boys sustained painful bites but had no ill affects. I don't think they will be trying to catch any snakes for a long time.

Skulduggery In The Suburbs

THERE ARE SOME ANIMAL catchers and exterminators that have cashed in on the average person's fear of snakes and lack of knowledge about these animals. There are many stories of families calling for a snake expert to come to their home to remove snakes and when this expert gets there he discovers—to the owner's horror—that the snakes are venomous and he will have to raise his fee to "exterminate" them. The family, of course, will pay anything to get rid of these deadly creatures.

This happened to a wealthy family who had nothing

more than garter snakes in their yard, but when they contacted the police, they were told to call an animal trapper listed in the phone book. When the animal trapper responded, he told the family that he was a "snake expert" and that the snakes were copperheads and very dangerous.

The parents were about to pay the man a fee for removing the fearsome creatures, but they happened to have two children attending an elementary school where I had just given a lecture, and the children suggested calling me.

When the snake expert learned I was coming, he took off, and I arrived to find that the snakes were harmless local garter snakes. He had frightened them, using their lack of knowledge about their environment to try and extract a handsome payment for himself.

I caught the snakes and relocated them in some woods nearby. The moral of the story: Know what kind of animals live in your area and be careful of "experts." The real snake may not be where you think it is. And it may walk with two legs.

Sometimes snakes and other critters make their way to your home by more unusual means. This happened to a family north of Dayton who went camping in Florida. When they returned home and opened one of their suitcases, they found a two-and-a-half foot yellow-colored snake staring at them from their underwear.

The husband immediately slammed the suitcase closed and called the police, who called me. Upon my arrival, I opened the suitcase slowly, not certain what kind of snake it really was. After seeing it, I identified it as a Florida rat snake, which is a nonvenomous snake that had apparently crawled into the luggage bag looking for a place to hide, not realizing that he was going to take a trip to Ohio. This kind of transportation of an animal happens quite often and many a family has been surprised by such hitchhikers.

Four-foot Creatures Masquerade As Nine-foot Creatures, Cause Fear Among Longtime Residents

THE LARGEST SNAKE CAUGHT in the Dayton area was a Burmese python over fifteen feet long, crawling loose near a restaurant in the downtown Dayton area. The snake was caught by firefighters at Station #11, who got the snake into a container and then had an animal control officer come to get it.

This is where I came into the picture, as animal control has nowhere to take the snake or anyone to take care of it. I found the snake a home with a Herpetological Society member from northern Ohio. The snake was a pet and had gotten loose from its owner. The animal control people thought it had been loose for a while, as it had regurgitated four baby opossums.

The sad part of this story is that Burmese pythons are fine when they are one to three feet long, but they can grow to be over eighteen feet long. Then they become expensive to feed, eating rabbits and large mammals, and are hard to handle. They can weigh as much as 175 pounds, and are no longer a logical pet, which is why some snakes escape or are turned loose by their owners.

I blame pet stores for not giving out proper literature about large snakes, and I also blame the buyers of these snakes for not studying and getting the information about their new pet before buying the snake. Out of the twenty exotic snakes I catch each year, ten are usually Burmese or reticulated pythons, the other five are usually boas or some form of exotic snake. Every time I am called out, I never know what I may find.

Most of the calls I receive about snakes usually involve local snakes, such as the time I was called to an assisted living facility where a large poisonous snake, said to be nine feet long, was not allowing people in or out of the facility, hanging by its tail at the main entrance.

Upon my arrival, the head of the facility rushed up to me with hope in his eyes that I was the snake guy the local police said was coming. I walked up to the snake and saw that it was a four-foot long black rat snake, which is common in our area and well-loved by the farming community for its superb rat-killing capabilities.

The snake was never aggressive, and I reached up and lifted him off the door area and began to show everyone what kind of snake it was. Even in this age of knowledge from books and television, people still kill nonvenomous snakes that are important to our environment, such as the much misunderstood black rat snake.

Many times I have responded to a call where some stalwart fellow has chopped the head off a black rat snake with a hoe, thinking it was some dangerous adversary.

In our area, there are many city folk moving to new homes built on the edge of heavily wooded areas. The snake crawls out of hibernation to discover that his territory has been invaded by large destructive creatures who weren't there when he went to sleep. He crawls up on the porch that also was not there when he went to sleep, and these large frightening creatures either try to kill him or they call the police, who may also try to kill him.

These new homeowners should go to the library or nature center and learn what is living in their area and how important these snakes are. An equation that should make people more hospitable to the black rat snake is this: would you rather have a few black rat snakes, or hundreds of rats?

Say It With Flour;
Monster Falls For Old Herpetologist Trick

THE FOLLOWING MAY SOUND like a tabloid title but it is not. "Monster anaconda that preyed on an infant" was a story on a local news station. Local animal control personnel called me and when I went to the home, I found the owner petrified with fear.

I soon learned that the property owner had two home improvement men working on the interior of the home, repairing damage from the family who had fled in fear from the snake.

The first worker said he saw the monster come crawling out of a hole in the wall above a baby cradle. The man showed me the hole that was the size of a saucer used for a tea service. I asked him what he saw, and as he spoke, I realized he was a master storyteller. This had happened three days ago, and he had many times to tell and embellish his story.

He said the snake had stuck his head out of the hole above the cradle, then slowly slithered out, all thirty feet of him. He had seen a show on anacondas, and he was absolutely sure this was an anaconda.

The worker would look at his partner and say, "That's what happened, right?" And the partner would shake his head in agreement.

After the two witnesses left, I checked the hole. It was obvious that a snake, described in the story as being thirty feet long, could not have crawled through this hole. The owner said he had taken a picture of the beast, and I became more excited, thinking perhaps there was an exotic snake loose in this home.

The owner handed me the picture, which was mostly the tail of what appeared to be a snake crawling into a hole. From the picture, the tail seemed to have the colors and patterns of a royal python. I estimated the snake to be about six feet long.

And then my travels came into play. Python hunters in India taught me a trick I thought I could transfer to Dayton. I placed a thin layer of flour on the floor near the hole. This should prove if there was a snake or not. When the snake came out of its hole, it would crawl across the floor, leaving a trail in the flour.

Then I devised a box with a hole close to the same size as the one in the wall, only a little smaller, just enough so the snake could crawl in and get his prize, a rabbit placed in the box. We left the box overnight and the next day we had a six-foot royal python with a huge bulge in his stomach.

Unable to crawl back through the hole in the box, the monster was trapped. He turned out to be nothing more than a non-aggressive, well-fed python of average size.

No one ever knew if a baby and the python had been in the house at the same time. The python probably belonged to the family that had rented the property before the new family moved in. This family had departed suddenly, it seemed, and for some reason the snake had been left behind. Perhaps it had just become too large to transport around.

This story, however, was not reported by the media, probably because no one seemed particularly at risk—unless it was snake—and the story was no longer sensational enough.

Every so often, however, I receive a call from a local police department stating that someone has reported capturing a poisonous snake and they are exactly right. Such is the story about the man who had moved to Dayton from Kentucky and was mowing his grass at his new residence when he discovered a snake.

The snake was obviously not a Dayton snake as it had a rattle on the end of its tail. The man covered the snake with a large pan and called the police. When I arrived, the man said he knew it was a timber rattlesnake because he had lived in Kentucky where they are found.

As I approached the pan, I looked at the man and I

could tell that he thought he was telling the truth. I lifted the pan and used my snake hook to lift up the snake.

There was no question that this was a timber rattlesnake, and the man was absolutely right. As I bagged the snake I thought to myself, *how did this snake get to Dayton?* It was also the question asked by the man and his neighbors.

The likely answer is that the snake was removed from its natural habitat, sold through a classified ad or at an animal swap meet, then turned loose when it became too much trouble to feed and properly care for.

Lawless Society Creates Breeding Ground For Exotic Animals

I HAVE WRITTEN ABOUT HOW PEOPLE obtain exotic pets, but some people cannot believe how easy it is to get a cobra or rattler. In the Dayton area, all you have to do is look in the classified ads of the local papers. There are no laws against having a deadly venomous snake, or a lion, or even an alligator within the city limits of most cities. Anyone can have a mamba or cobra in his apartment—*your* apartment complex—that can escape or be turned loose.

Most emergency room doctors do not know how to treat a venomous snake bite and in your city or nearby zoo there is usually not enough antivenin, if any, for the specific snake that has bitten you.

My son, Adam, was 14 years old when he went to an exotic animal swap meet at a local fairground where we observed a man selling Indian Cobras to the public. I told Adam to go talk to this guy and see how far he could get in purchasing a cobra. Adam was at the exhibit three minutes when he had bought himself a cobra. If a 14-year-old boy can buy one, who knows what your son or daughter might have in his or her room, or what your neighbors might have as a pet.

The last venomous snake bite in which I was involved was by a prairie rattlesnake that the victim bought through the *Trading Post* classified ads. After fortifying himself with liquid courage, he decided he would handle it as in the *Crocodile Hunter*, a popular cable show where an expert from Australia allows venomous snakes to crawl all over him without getting bit.

This man was definitely not thinking clearly, and the snake bit him on the forearm and he had a full envenomation. He almost died, but the hospital acquired antivenin serum from my brother. The serum has to be given within 24 hours, but it is dangerous, as well, because it is made from horse serum, to which forty percent of the population is allergic. An allergic reaction can be as bad as the snake bite.

Healthy people, without previous allergic reactions, almost never die after being bitten by a rattlesnake or copperhead, even without the antivenin serum. Blood blisters appear, which look like gangrene, and they disappear after about three days. Sometimes, there is a little scar tissue from the bite itself. Thus the victim does not die, although during the three days he or she may *wish* to die.

Until we have some common sense laws to protect people from themselves, as well as the public, we can only be aware that the majority of venomous snake bites occur when someone tries to kill one, catch one, or show off with one.

First, the snake does not want to bite you. In my children's program, I have a child lie flat on his or her stomach and I walk up and stand over them. I ask them how it feels to see someone standing over them. Everyone says it is scary. Now, imagine yourself as a snake. How scary would it be to have only a head, no hands or legs, and a human looming above them?

In our lectures, my brother and I use the story of two men who went walking into a Kentucky forest with their handguns as protection. One of the men spotted a rattlesnake and attempted to shoot the rattlesnake. Instead of shooting the rattler, the bullet ricocheted off a rock and struck his partner in the abdomen.

The man fell to the ground, landed on the rattler, and was bitten on the rear. His buddy tried to help by pushing him off the rattler and, in the process, was bitten on the hand. Now there were two snakebite victims and a serious gunshot injury, all over a snake they could have easily walked away from and left alone.

It should be noted that there is no snake that crawls faster than a man can walk, and I have never heard of a snake attacking a car or truck or jumping through a window after a passenger or driver.

Maybe the man in the Kentucky forest should have simply shot his partner, rather than firing at the snake. That way, there would have been only a gunshot wound...

To prove just how much people hate snakes, we placed a rubber snake on a country road and watched how people would react. We had cars and trucks purposely run over the rubber snake, then stop and back up to run over it again. Some would get out of their vehicles and hack the rubber snake with machetes, garden tools, tire irons, and whatever else they might get their hands on.

When they finally noticed that the snake was rubber, they looked rather sheepishly around to see if anyone was watching. Then they quickly drove away, a little trail of embarrassment in the air.

People ask me how to keep snakes out of their yards and homes and my remedy is easy. Snakes only go where food is. If you have wood stacked in your yard, out buildings not cared for, high grass or weeds— anything that would draw mice or rats—then you should be thankful you have snakes.

If you keep your yard area clear and clean, you will not have the food source and the snakes will not be around.

It's that simple.

Lunch For A Burmese Python, And Other Stories

I WAS ONCE CALLED TO A HOME where a mother wanted me to look at a new pet snake that her son, who was 10 years old, had just bought at an exotic animal swap meet. The snake was not eating. I was taken to the boy's bedroom to see the snake and when I looked into the aquarium, I was shocked. The mom had been trying to feed small pinkies—baby mice—with her fingers, to the snake and the snake was lethargic.

I asked her if she had ever heard a little poem that people say in the south, "Red to yellow, kill a fellow, red to black venom lack."

"Yes," she said and acted puzzled.

I asked her to look at her son's snake and tell me the color pattern. When she started to recite the poem, an ashen color appeared on her face. She became aware that it was a coral snake, highly venomous.

Sometimes parents do not know what kind of pet their kids have brought home, sold by some irresponsible animal dealer.

In north Dayton, a man called the local police to tell them a snake was in his garage. This garage was attached to four other garages and belonged to an apartment building. His two children, 4 and 6 years old, were found playing with the snake.

When I arrived, I walked back to the garage and met the two boys. They walked me into the garage where the oldest boy moved a trash can from the wall and pulled a hissing, three-foot snake out by its tail.

I was shocked by what I saw, for the snake so unceremoniously yanked out by the six-year-old boy was a Gaboon viper, a deadly venomous snake from Africa. I told him to let go of the snake and move away from it.

I caught the snake and showed the children and their father the Gaboon viper's fangs, which are considered the world's biggest. Everyone was stunned. This incident taught me to never take any call for granted.

The kids were extremely fortunate. The only reason they weren't bitten was because the viper was lethargic from being in such poor condition and didn't have the strength to bite.

When a snake, especially an exotic snake, bites a human it always makes the news. Once, I received a phone call from a frantic police officer about a 14-foot Burmese python that had grabbed a young man's hand in its mouth and wrapped itself around the victim. Everyone thought the snake was trying to eat the man.

I told the officer to call the ambulance and pour hot water on the snake or rub ammonia on the area of its nose; both will cause the snake to release its prey. There were two police cruisers and a medic crew already there when I arrived. Entering the apartment, I heard bloodcurdling screams from an upstairs bedroom. I raced up the stairs to see paramedics trying to exit the bedroom with a man bleeding profusely from his hand.

The paramedic's eyes were full of fear as he yelled that the snake was loose and trying to bite everyone. The police officer flew out of the bedroom and sprinted by me without an acknowledgment. A large Burmese python was curled up in a corner of the bedroom.

As I approached, the snake struck at me and almost grabbed my right thigh. When I jumped to the side, the snake recoiled itself and struck again. Bent over and off-balance from my leap to safety, the snake struck at my face. Its speed was impressive.

I leaned back just enough to save my face, and the snake hit me square in the chest, knocking me backwards and forcing the wind out of my body. When it recoiled to strike again, I rolled out of its striking distance and regrouped.

When I regained my senses, I caused the snake to strike again but this time I grabbed it by the neck, securing the gaping jaws from biting me and yelling for help to control the snake's body.

The police officer had called for backup, and they rushed up to my assistance. The officers secured the snake's body, and we placed it in a holding box and moved it to my vehicle. Now that the snake was secure and the scene was safe, I checked the original victim, who was with the medic and still shouting in pain.

His hand was so damaged, tendons were exposed and most of the flesh ripped away. I asked the police officer what happened and he said the victim's roommate told them he could remove the snake from his friends hand, and he grabbed the snake by the head and yanked as hard as he could, ripping the snake away but taking most of the flesh and some of his friend's tendons in the process.

Pythons have substantial teeth used to hold their prey so they cannot escape. The teeth are curved back towards the throat area and are extremely sharp. When the roommate ripped the snake away, he permanently damaged his friend's hand.

The snake was over fourteen feet long and weighed over 100 pounds. The police officer said he did not have the time to use my remedy because of the roommate's enthusiastic assistance. I told him it was not his fault.

Snake Fastens Itself To Woman; Mistaken Case Of Fatal Attraction

WHEN I WAS IN CAPETOWN, South Africa, filming a show on great white sharks, an American friend brought me a newspaper—South Africa's equivalent of the *National Enquirer*—and one of the stories was about a python attempting to eat a woman. What was unique about this story was that it happened in Dayton, Ohio.

The article said that a python turned on its owner, grabbed her by the hand, and wrapped around her body. Then the snake attempted to eat her by swallowing her hand. "A python could easily swallow a young woman," said the expert. The paramedics and police arrived and heroically cut off the snake's head, which caused it to bite harder and now could not be removed.

The woman was transported to a local hospital where the snake's head had to be surgically removed from her hand. This story, appearing in papers all over the world, exemplifies the hysteria and misinformation that the general public and professionals have about large snakes.

The true story of the snake attack was that the woman had been playing with a pet rat and let the rodent crawl all over hands and arms. After she had put the pet rat away, a friend arrived and asked to see the snake, which had not been fed for a few months.

Snakes can eat a large meal, then they curl up and digest what they have eaten, which takes quite a while because they digest everything. Their metabolism is slow, and they expend little energy over considerable periods of time.

When the woman reached in to pick up her snake, the snake tasted her hands with its tongue and immediately struck at what the snake thought was a rat or a mammal. It then wrapped itself around what it thought was a rat and attempted to eat her hand.

The python was only about eight feet long and too small to swallow her hand, certainly too small to swallow the woman.

The moral of the story is: Always wash hands before handling snake, to prevent problematic cases of mistaken identity.

Make sure and wash one's hands after handling any reptile, which helps prevent any salmonella problems. If a snake strikes and bites you or another person, always remain calm; do not pull or rip the snake off the victim.

Second, control the snake, since a large snake will choke a victim unconscious or the snake's body weight may cause the victim to fall. Third, pour or place the snake's body in hot water, which usually causes the snake to let go of its prey. Fourth, if you have ammonia—or better yet, ammonia capsules—rub ammonia on the snake's nose area. This technique almost always works. Stabbing the snake or cutting off the head can cause more pain and damage, permanently injuring the victim and killing the snake.

Because of the medical field's lack of knowledge about venomous snakes, especially exotic snakes, the treatment in an emergency room might actually be worse than the bite. In some cases, treatment itself can cause permanent damage, even death.

There are still medical personnel that believe in the cut, suck, and spit method. This method is outdated, ineffective, and has caused many a snakebite victim to have a permanently crippled extremity after an excited first-aid rescuer lacerates tendons in the victim's hand.

I have seen many a "rattlesnake roundup" competition where the media interviews a past "snakebite survivor," who shows his damaged, claw-like hand to the television cameras and tells how a rattler did this to his hand.

In many such incidents, the damage is done not by the snake but by the hero's buddies, who, after their friend has been bitten, pull out their knives and perform an incompetent surgery.

In any case, the hand is a delicate part of the human body. It is easy to damage the intricate tendons and nerves of the hand, so why would anybody want Billy Bob (who most likely has had a few beers) take out a knife or razor and cut into such an important extremity?

Most venomous snakebites are dry bites or partial envenomizations and need no surgery or cutting. Every time I see these "survivors" on television showing what the demon snake did to his hand, I understand why the truth is seldom reported: No one wants to hear that the person essentially injured himself. The snake was little more than an innocent bystander.

My brother, Jim, once did a commercial for protective chaps to be worn by hunters and outdoorsmen to prevent snakebites. Jim tried to get rattlesnakes to strike but they just sounded off with one of the planet's best early warning systems—the rattle—and kept themselves at a safe distance. After many tries, he finally got one of the rattlers to strike.

How dangerous these creatures are, he laughed. First, they warn you by making all this noise. Then, they have to be tormented into striking. Which illustrates how truly dangerous a venomous snake is, if merely left alone.

A final story about a truly monstrous creature: A snake in a bag had been dropped off at a local nature center. Inside was a western diamondback rattlesnake, extremely thin and lethargic. The naturalist called me to check out the snake and try to find it a home.

When I first pulled the snake out onto the floor of the center, I could tell that this rattler had not been taken care of. When I pinned its head and picked up the snake, I discovered that the poor creature's rattle and base of its tail had been cut off.

It appeared that the snake had been tortured by its past owner, most likely so that this preening cowboy could have a rattle to put on his hat or for some other display, making, no doubt, a statement about his legendary testosterone count.

I wrote earlier of a truly monstrous creature. The creature is not the snake, but man himself.

t•w•o

Lions Tigers & Bears, Oh My!

Hairy Hitchhiker Startles Normally Placid Neighborhood

BACK IN THE EARLY 1980S, African lions and Bengal tigers were very common in the classified ads of local newspapers, and people kept them as pets. As cubs, they are adorable and easily manageable. They are less adorable when these cuddly babies outweigh their owners. They may become dangerous and are either forced into small cages or suddenly they escape and become a danger to the surrounding neighborhood and themselves.

When this happens, I get a call. Once, a large African lioness was walking down a suburban street, causing parents to grab their children and lock their doors. The authorities were trailing along behind the lioness in their vehicles, trying to decide whether or not to dart the animal with anesthetic.

At the moment, though, the lioness was merely out for a stroll, not being threatening at all, and the veterinarian said that if the anesthetic was administered in too small a dose, the lioness could become dangerous.

No one wanted to shoot her, both because she was not threatening and also because should the shots miss, there were numerous homes nearby to stop the passage of bullets.

The first thing I noticed about the lioness was that

she was walking bowlegged and looked malnourished. I pulled my car up next to her, looked her in the face, and saw she had cataracts that made it difficult for her to see.

I decided to try something that had worked before with such animals: I opened the back door of my vehicle to see if she would like a ride. When I opened the door, she stumbled over toward my voice. Through her cataracts, she could see an open door and hear someone saying, "Want to go for a ride?"

She put her front paws on the riding board alongside the trooper but could not raise her hind legs up to get into the back seat. I walked over and lifted up her hips and legs and gently pushed her into the back seat.

She appeared comfortable, and I turned and explained to the crowd that most lions kept as pets have had their front paws de-clawed, and they have usually been spayed or neutered. This lioness had, as well, assorted illnesses and infirmities of age.

There are no laws or ordinances against people having exotic pets in the Dayton area, which many times means that the danger to life and public safety is not the animal but the incompetent animal owner. In this case, the elderly lioness was returned to her home. I felt sorry for her and never heard anything about her again.

Another time, I was called by a local police officer who said his department was about to raid a house on the west side of Dayton that was a haven for motorcycle gangs. The raid was for drugs, illegal guns, and contraband, and he thought I might be interested because they were supposed to have a lion cub on the property.

I told him to contact me if they needed me. The raid went down two days later, and I was called to remove the cub. Behind the house, in a barn used as a garage, the owner of the property had pit bulls and a young lioness. When I approached the lioness cub, I saw that she was tied to a stake in the ground and had trouble breathing. Mucus was coming out of her nose and eyes.

The cub was about four months old and looked malnourished. She was lethargic but she tried to pull away from me as I picked her up and placed her in a portable cage in my vehicle.

On the way home, I stopped to see my veterinarian who said she didn't look like she would make it through the night. He injected her with her antibiotics and cleaned up her eyes and nose. She lay there barely breathing, and when she did breathe, mucus appeared in her nose.

I picked her up and brought her to my home where I gently placed her on some blankets next to my bed and stayed up all night watching her. During the night she stopped breathing, and I revived her by blowing down her nose and mouth, trying to keep her stimulated.

This went on all night and into the morning. By afternoon, the virus broke and she began to try and stand and wanted water. By early evening, she had eaten some meat and began to respond when I spoke to her. Her eyes and ears would perk up when I talked to her. She looked as though she would live.

Out of all the animals I have rescued, this little lioness would become one of my favorites. As the days passed, she began to eat and grow stronger. We became very close.

She would stalk me and play as if we were a pride. On the days I was working 24-hour shifts, my girlfriend came over and fed her.

One day she went to feed the lioness, which we began calling "Tabitha," put the food on an unbreakable food container, and walked down some steps to the basement area where Tabitha stayed when I was working. Tabitha liked to hide under the steps, and when the innocent person walked down the steps, she would reach from behind the steps, grab your leg, and trip you.

I had forgotten to include my girlfriend in our little game, and she expected to walk down the steps, put the food plate down, and leave. At this time, Tabitha was close to 150 pounds and quite playful. She looked through the steps, waited for Patty to step down on the steps, and she reached out and grabbed her ankle. Patty tripped and—worse—tore the new pair of pants she was wearing. Patty yelled and smacked Tabitha on the head with the unbreakable plate, shattering it and throwing food everywhere.

Tabitha looked at her as if to wonder about the new game Patty was playing. The plate had no effect on Tabitha. She continued to play and Patty left. I received a phone call from Patty and an angry voice told me I owed her a new pair of pants.

When I began to laugh, Patty slammed the phone down in my ear. I bought Patty a new pair of pants, and she paid me back by waiting until one day when I was in the bathtub, relaxing under a massive quantity of bubbles.

Patty brought Tabitha up into the living room area

and said, "Go get Daddy." Tabitha saw my head sticking out of all those bubbles and she sprinted across the wooden floor for the bathroom.

By the time I noticed her, she had leaped six feet across the bathroom floor, gone airborne, and landed with an immense splash on top of me in the tub.

Water splashed everywhere and she freaked out. Tabitha hated the water and she let out a deafening roar. I grabbed her and held her in the water until she was soaked and when I finally let her go, she slipped and slid across the wooden floor, howling and roaring, looking for my bedroom to hide in and lick off her wet fur.

Patty was roaring with laughter, and so was I. Tabitha was close to 200 pounds then and she crawled under my bed and would not come out. She pouted and was mad at us for the rest of the evening, but by bedtime she crawled into bed with me for her nightly chewing session, and all was forgiven.

Tabitha was raised differently from most lions. When she was young, I started mouthing her just as I did with my wolves, bears, and other large carnivores. By this I mean I would roll her over on the ground and chew on her neck, arms, and face. She would do this with me and she learned how to be very gentle. I rarely used voice commands; I just tightened my arm if she bit too hard and she learned to ease off just by my tightening. Mouthing made us connect and we became close. I could read her moods, and she could read mine without any words being used.

I never train my animals, rather I try to get an

understanding between us so that I know what they need or think and they know what I expect—a mutual trust. If my cat is acting nervous, I want her to look to me for comfort and I always help her to feel safe.

Tabitha grew to be a beautiful 450-pound lioness, and she had outgrown my home. She needed playmates her own size. All my big cats had their claws and were exactly how our heavenly father created them. A large percentage of big cats in the care of individuals have had their claws removed and have been spayed or neutered.

These cats can no longer be placed with zoo or animal park big cats because they would be unable to defend themselves. People buy a cute cub, thinking that a zoo will take the cat when it becomes uncontrollable and dangerous. They do not understand that zoos do not want other peoples' mistakes. They do not want animals that have not been fed properly and may have medical problems.

These people may honestly love their animals yet find themselves in a position where the choices are euthanasia or selling them to individuals who use them for big game hunting farms. Some people love animals to death.

In my case, I found a zoo in Nashua, New Hampshire that was looking for a controllable, healthy lioness. I drove out to the zoo and saw that they had a beautiful, spacious area for their lion exhibit, and they had a young male around Tabitha's age and size.

I immediately decided this was Tabitha's new home and brought her to New Hampshire to live. She was quarantined for a period of time, then placed in with the male, and they got along fantastically.

The funny part of the story is that we transported Tabitha by station wagon. While driving, my wife, my cousin Cindy, and I, all took turns lying in the back with Tabitha to comfort her and keep her company. Lions are animals that like to be in a family environment, such as their pride. They are much like wolves and their pride is much like the wolf's pack.

Tabitha considered us her pride; she liked to be around us and followed me like a dog. She always liked one of us near her and felt comfort from our presence. While driving through Massachusetts, a state trooper stopped me. As the trooper approached my vehicle, Tabitha woke up and watched him approach. The trooper asked for my driver's license, and as I reached for my license, Tabitha stuck her large head out the driver's window and into the troopers face.

The trooper stumbled back a few feet, his eyes the size of saucers. At first he could not speak, and when a few syllables did come out of his mouth he spurted, "Li-Li-Li-Li- Lion." I said, yes, he was correct.

He handed back my license from a great distance, stretching every inch of his body towards my hand. I asked him if I could go and he shouted as he headed back to his cruiser, "YES!"

I do not want people to think that you can get out of a ticket driving around with a lion in the back seat, but it worked in Massachusetts. I phoned the zoo in New Hampshire a few years after we dropped Tabitha off and learned that she had cubs since I last saw her and was doing great.

Tabitha was my favorite big cat. She nearly died, never destined to have been a pet, and by a fortunate accident became a splendid creature. I will never forget her.

Man-eater Disappears After Sowing Fear In Township

WHILE WATCHING A LOCAL NEWSCAST, I saw a story about a Bengal tiger walking freely through a township south of Dayton. A man working in an office building had looked out the window and was startled to see a tiger walking across an open field. There were wooded areas surrounding the office building and homes framed the open field where the tiger was spotted.

When the news media found out, the incident immediately turned into a circus. Various experts from a nearby zoo came in and caused panic to the surrounding communities by making statements such as: "Tigers are known man-eaters." (Apparently, women and children were safe if the tiger could only find a man.) I thought no one should have known that a tiger was on the loose until the tiger was verified to exist. I came out to the location on the second day and found no footprints or any other signs of a tiger.

The experts spent three weeks looking for this tiger. They used helicopters, four- wheel drive vehicles, infrared binoculars, and bait, all to no avail. They gave news conferences and enjoyed the publicity.

After three weeks of this fiasco, the search was called

off and the experts, not seeing anything, condemned the witness—even though he had a video of the tiger. The experts said it was a tabby cat seen from a distance. Earlier, the same people had seen the same video and proclaimed it was a tiger.

When boredom set in, all menace evaporated, and in the manner of storms losing intensity, the tiger was downgraded to a house cat.

Upon further investigation, I discovered that a local man owned the tiger. About a year before, the man had been in trouble with local authorities about his tiger escaping and causing a panic. The escape was broadcast on all the local television stations and newspapers.

For some reason, no one seemed to recall the previous incident. And the media made no connection between the two.

No one questioned him this time, or even asked him the whereabouts of his tiger. Through my sources, I was able to locate him. My theory was that his tiger may have gotten out and was immediately retrieved by him a short time after the escape. He laughed off the theory, although he did not deny it, admitting that even he found it peculiar that no one thought of him and his past history of tiger problems.

The officers remembered him, though. They knew he had kept his tiger in a hay-filled trailer behind his girlfriend's house, and that he had gotten out, crawled into the back of a police cruiser and chewed up the upholstery. Once, the guy was keeping the tiger in a barn, which caught on fire, and the tiger rushed out of the burning barn, past startled firefighters who couldn't believe what they were seeing.

The poor tiger had lived all over the county, moved here and there by its ne'r-do-well owner, who failed to properly care for it. Like this man, most people think tigers can be fed chickens, but tigers need more—vitamins and minerals, good red meat, bones to crunch on, blood—a nice deer, for instance.

Only a few people knew the true story of the tiger hunt near Dayton, and even fewer wanted to believe it.

A Stalk On The Wild Side

I ONCE HAD AN OPPORTUNITY to receive a three-week-old Bengal tiger cub from a zoo in the northern part of the state. The cub was the runt of the litter. This cub had gastric parasites and no veterinarian care. It cost hundreds of dollars just to get her healthy.

People do not realize how expensive it is to take care of a tiger or lion, which are high maintenance animals. I have not yet even mentioned the time it takes to adequately clean and care for them. This is why I try my best to dissuade people from having these animals as pets. I want them to understand the financial, emotional, and physical toll a pet like this takes on its owner.

My little cub grew to be a majestic 400-pound tiger that was affectionate and extremely controllable. She was named "Nicki" and I used her for my school programs or when groups would come to my compound area to learn about tigers.

I would have the naturalists and interns from a local nature center come out, all of them having an opportunity to interact with "Nicki," and feel her fur, teeth, claws, muscles, and her rough tongue. Such interaction with the alleged man-eater showed people that tigers are not savage, uncontrollable killers, but only an animal that does what it was created to do.

In India, children understand that if they walk along a path and a Chital deer is on the same path, a tiger, looking for something to eat, will choose to attack the weakest, slowest, and easiest to kill. Humans fit all these categories. In India, they understand this. In the United States, we still think that a large predator should understand the theory of Manifest Destiny and that we Americans are the chosen species.

People have the "Bambi" syndrome—animals think and act like humans and will never harm us because we love them. I relate a sad story to my school groups, one that happened in a national park in the western part of the United States. A mother poured honey over her infant child's head so a black bear could lick the honey off while she videotaped this cute interaction. It was an example of the tendency that people have of leaving their brains at the border of national parks. They believe that they will be able to walk up to a bison or bear and not get hurt.

The black bear began to lick the honey. It was not aware that the object it was licking was an infant, and crushed the child's head, killing the child. Common sense would tell you that the mother should have been arrested, but that was not the case. For the bear's great error in judgment, it was shot and killed by rangers.

A hungry tiger or bear will see only a slow, weak, easy-to-kill creature. A child of any age is slower and less of a challenge than a Chital deer that can run up to forty miles an hour. People should be aware of these facts and understand that if they enter a jungle, woods, desert, or body of water, they are entering something else's home. Act accordingly. Learn about the animals in these areas and what to expect.

Tigers are more catlike than lions. This means that tigers can be happy alone and not need other tigers, or us. They do not need a pride or pack to survive. They think independently and aren't as needy. Nikki was very affectionate, but only when she wanted to be, whereas Tabitha, even if she was not in the mood for a scratching, would immediately allow me to scratch her and respond with a mutual affection.

Nicki, if she did not want to be scratched, would get up and move away and not allow you to pet her. She would only allow you to pet her when *she* wanted you to pet her. There would be days when she would continually rub up against me and want me to chew on her neck, but there were days she did not want my contact and I always left that up to her.

Reading your animal's body language is important. If you are dealing with a large predator, it could save your life. Leopards, for instance, are one-person animals. I have met a few both in captivity and in the wild. In Nepal we were tracking tigers and the ground was muddy. We left our footprints in the mud and when we walked back to our camps, following our tracks, we observed leopard tracks inside our footprints.

The leopard had been tracking us while we were tracking the tiger. He was merely curious, walking along and smelling us, tasting our tracks. After all, he was curious and like most of mankind itself, looking for something easy to eat (which explains the popularity of fast-food restaurants). When he discovered that we were not particularly easy, he didn't bother us.

The lesson: No matter where you are in nature, something is watching you. You must pay attention to the language of the forest and jungle, because you are in something else's environment. You are no longer in control. We stalked the tiger, and we, in turn, were stalked.

I have a friend who had a leopard as a pet. One day a friend of *his* came to the house, walked in the front door, and approached my friend. The leopard attacked the man, grabbed his throat, and killed him almost immediately. The owner had never seen any aggression from his leopard before.

In my travels, I learned that if a leopard perceives a threat, he will act aggressively, even though we may not understand his actions. The cat saw something we did not, and that behavior should deter people from having these big cats as pets. No one knows what might scare them or put them into an attack mode.

Nicki had a fear of lawnmowers and would run through a wall to get away from one. She also did not like it when I put on cologne. She would growl and grimace every time she came next to me. She was a rare tiger in that she loved to take naps with my boys and me, lying on top of us like a big furry quilt. She would wake us up by licking our heads with her rough tongue. I think that's why I'm now bald...

My oldest boy, Adam, was six at the time and told everyone that he did not have a cowlick but a tigerlick. When Nicki became too big for us, I began to look, nationally, for a zoo where she could be with other tigers. I traveled around to many zoos and did not like any of them.

At that time, most zoos did not spend much money on animal care or veterinarian services unless they were for money-making animals like white tigers or condors. Some zoos were selling their excess lions, tigers, leopards, and bears, along with hoofed animals, to large hunting ranches in the western part of the United States.

People would pay big bucks to hunt a large cat on one of these fenced-in ranches. The majority of cats had cataracts or were elderly. The ranchers would release a tiger that had been fed by man, raised until it was too old and feeble to be on exhibit, then allow a hunter sitting on the back of a pickup truck to shoot it. After the cat was killed, the hunter would take it to a taxidermist. The taxidermist would work his magic and when he was done, the cat would look like a young man-eater. That was the racket.

I also had to worry about zoo personnel wanting a fertile young female tiger to breathe fresh air into the breeding of their white tigers. They were always looking for new blood, and I almost fell for a zoo that wanted Nicki to breed with its white male tiger.

The problem was that these professionals were not telling me the downside of my tiger breeding with theirs, and they romanced me more heavily than the courtship of the male tiger itself.

Luckily, I knew a past zoo professional who had worked with this male tiger, and she said that the last two females put in his cage were mauled and killed. One of them was eaten. This information was suspiciously left out of our meetings with zoo personnel. I became aware that I could not trust anybody.

I finally found a zoo that had a new area for their tigers, and it had a young male that I actually had a chance to interact with. I brought Nicki over to the zoo and she hit it off with the male and actually acted as the dominant cat. I found a home for her, and she seemed to be very happy.

Financially, I spent hundreds of dollars on her. Physically, I had numerous scratches and bruises from aggressive play. The worst part, however, was the emotional part. You become very attached raising a big cat, and leaving them is very difficult. I know she is happy and where she should be, but I will miss her, and that is the hardest part.

A story I tell about how big cats I have raised look to me as their human tiger is best illustrated by an incident where I was standing on stage at an elementary school with Nicki. The children entered the auditorium like water pouring through the doors. I had already told the teachers to make sure the children came into the auditorium calmly and one at a time, so as not to startle Nicki.

But they all boisterously charged the stage. I yelled

on the microphone for the teachers to take control of their students, but they laughed and stood back.

Nicki went crazy with fear. She was well over 200 pounds at that time. I will always remember her looking at me, her eyes full of fear. She looked to me for help and comfort, so I pulled her leash and I headed for an exit on the stage. I went straight into the parking lot, got into my vehicle, and left the school.

I received a nasty call from the principal about my hurried exit. He wanted his money back for the show. I explained what happened and that my animal's feelings come first and that he would *not* get his money back.

This story explains how people disrespect animals, even educated teachers who should know better. They should have been aware of how their unruly students could have caused Nicki to hurt one of their students. Nicki trusted me as a safe zone. She knew I would think of her first and not leave her in a scary situation where she would have to act aggressively.

When a big cat in captivity hurts or kills someone, usually someone is showing off and forcing the cat to be where it does not want to be or to do something it does not want to do. Innocent people are hurt when big cat trainers allow the cat to become tired and agitated or force it into situations where people surround the cat and make it nervous or walk up behind the cat and surprise it. The newspapers then run a story entitled, "Tiger Turns on Owner." or, "Victim Tried to Pet Tiger, Seriously Mauled."

The truth is that the owner turned on the animal by forcing it to be or to do something it did not want to do.

There are some big cats that are emotionally and psychologically damaged and it can be difficult to predict their behavior. Common sense would tell their owners to be extra careful in dealing with them. These are the animals that allegedly escape, or are turned loose by the owners. I have yet to figure out who is more dangerous—the big cats or the owners.

Garlic Sauce Causes Sensation In Condo; Diner Shows Appreciation By Going Through Wall

THE LAST ANIMAL IN THIS CHAPTER is an animal that was very popular as a pet in the 1970s—the bear. The bear is an animal that is misunderstood and mistreated because of a lack of knowledge about what bears do naturally.

My first bear experience was with a black bear that a local farmer had as a pet, which had broken loose from a neck chain. The sheriff's office was notified of a bear walking along a country road and wandering into traffic. I was called to the area and quickly found paw prints along the side of the road.

Following the tracks, I located the bear curled up and sleeping near a boat behind a barn, several miles away from his home. The bear was a black bear and was about three years old. It was in bad condition and had a deep wound around his neck from the chain.

I slowly placed a rope around the bear's neck and gently woke him. The bear stood up and began to walk with me to the truck. We got into the truck bed, and the bear and I were transported to a veterinarian I worked for.

The bear was very docile, almost appreciative of our assisting him. The bear's owner wanted his bear back but the veterinarian, who did not want to give the bear back for more mistreatment, said the owner would have to pay the veterinarian bill first.

The farmer did not want to pay the bill so the veterinarian and I became owners of a bear. We went about healing this bear and we found an animal park that would take him. In taking care of this bear, I became aware of just how strong these creatures are, both from a standpoint of strength and body odor. I also learned that bears like to dig holes under fences and escape.

People do not understand how potentially dangerous they can be. They look cuddly, especially when they are young, but they can be ornery and their exceptional speed at quick bursts can surprise the fastest man.

The bear, which we called "Stinky" because of his pungent odor, healed and put on about fifty pounds of body weight. Stinky was transported to the animal park where he lived a happier life than one on the end of a heavy chain.

The most unusual place from where I helped rescue a bear was an apartment. Here, a bear lover had raised and was keeping a bear on the third floor. Neighbors heard growls, but they thought the sounds were made by something like a rottweiler.

One day the bear lover left to go to work and left the black bear, Cuddles, in the apartment alone. Despite his name, Cuddles was over 200 pounds and as the day went on, Cuddles apparently became hungry.

I know this because the neighbor in the adjacent apartment was cooking her lunch—spaghetti with heavy garlic sauce. The strong smell must have drifted from her apartment to Cuddles, causing him to try and locate the nearest Italian restaurant.

To accomplish the task of discovering the smell's origin, Cuddles began to rip and tear down the drywall. He proceeded right into the neighbor's apartment, tearing down her wall, and finally devouring the spaghetti.

I led the bear out with the spaghetti pan to the animal control van. She finished her dinner there.

The Bear Lady And Her Bear; A Love Story

YEARS AGO, I OWNED A KENNEL where I boarded dogs and cats. I was known for boarding exotic animals such as macaws, monkeys, snakes, cougars, wolves, and other exotic clientele. One unusually quiet day, I received a phone call from a friendly lady who wanted to know if I boarded bears. I told her yes, thinking her bear was possibly a black bear. She arrived by truck with a large trailer in tow. On the side of the trailer was painted a woman standing on top of an enormous polar bear with two black bears at their side. The trailer was part of a circus show that was being held at a local arena that weekend.

A ravishing woman of advanced years exited from the truck and approached me with her hand out. As I grabbed her hand, I felt a firm grip of an incredibly strong woman. Counting my fingers, I listened as she told me she was with the circus, and that she was the Bear Lady on the side of the trailer.

She said she had heard that I was good with animals, and that she had a special animal she needed to show me. The lady opened the back of the trailer and an enormous white figure emerged. I was amazed at the size of this creature. When it came into the light, I saw it was a polar bear, apparently the same one that was painted on the side of the trailer.

The bear moved slowly, as if it had arthritis. I looked into his eyes and saw cataracts. And as I was looking into his eyes, the bear opened his mouth and mouthed the top of my head. The bear's spit oozed all over my head and down my cheeks and neck.

He was extremely gentle and as he let go of me I noticed that he had no front teeth and only a few molars. This was one geriatric bear.

The lady said that the bear, whose name was Arctic, liked me, and was in need of help. Arctic was a circus bear who was on his last legs and needed a place to retire and finally die.

He was about thirty years old, had cataracts, no teeth,

and was overweight. He may have weighed 1,000 pounds or more, but he looked like he weighed two tons. He was very sweet and playful for his size and age, wrestling gently with me.

I placed him in a twenty-foot by fifty-foot, eight-foot tall, totally enclosed cage. Arctic seemed to like his cage and made himself right at home.

The circus lady said she would be back at the end of the circus run to check on Arctic and say good-bye. Arctic and I became close and I enjoyed playing with this extra large teddy bear. However, a problem arose on the second day when Arctic stood up and placed his gigantic paws on the side of the fenced-in cage, pushing it loose from the support poles and sending it crashing to the ground.

I understood that I did not have a cage or structure strong enough to hold this giant. I put a rope around Arctic's neck and led him to a building where I had dog runs and indoor cages. The dogs in the cages and the kennel runs went crazy seeing this bear walking by them. Some barked and growled while others cowered in the farthest corners of their cages. I got Arctic to lie down and he fell asleep.

The next day the bear lady came back and I explained how I did not have any cages in which to keep Arctic safely. She understood and made a decision to look for other places for Arctic to retire. Together we made some calls to zoos and animal parks trying to find one that could care for this big baby.

We finally found an animal park out west that had grass so Arctic could feel grass on its claws. There was shade and a comfortable environment. They took the bear and gave him a good home for his twilight years.

The bear lady thanked me for my assistance and I thanked her for the opportunity to have a few days with a polar bear. I considered this experience my ultimate bear encounter.

Here are my final notes about lions, tigers and bears as household pets. I have never heard a happy ending to a situation where someone has raised a big cat or bear in his or her home. The animal is usually euthanized or it escapes, which means that it was turned loose. Some animals cause their owners huge financial heartaches, as they may easily frighten or hurt someone. Lawsuits abound from these occurrences. If you like big cats or bears, volunteer at zoos or animal parks where you may enjoy the animals without retribution.

t•h•r•e•e

I Thought I Saw An Alligator

The Alligator Dude

I LIVE IN THE STATE OF OHIO, hundreds of miles away from any alligators, but over the years I have met Ohio alligators in local rivers, ponds, pet stores, back yards, and basements. You may wonder why there are so many alligators loose in Ohio. The reason for this phenomenon was the removal of the alligator from the endangered species list. Pet stores and private dealers found there was money in selling cute, foot-long alligators as pets to unknowing customers.

Customers were (and still are) naive because they had no idea what happened to cute foot-long alligators—they grow a foot a year. Pet owners suddenly had a large, aggressive alligator that was not easy to handle and becoming more expensive to feed by the day. Cute was out the window.

The owners no longer wanted this pet, and many thought, "I'll turn this alligator loose in a pond or river." At this point, the owner began to congratulate himself for doing a good thing. The owner was having what I call a "Born Free Experience."

There is always stupidity enough to go around, but the essential culprits here are the pet store owners. They are the ones who sold animals without informing the buyers that an animal is going to become a large predator that cannot survive in Ohio's cold winters. The buyers, of course, have not thought past their initial impulse, and haven't the slightest idea what lies (or crawls) ahead.

The only innocent character in the scenario is the alligator, who, if given a choice, would rather live in a swamp in Georgia than a bathtub in Ohio. At least, I think so.

One alligator incident began with a phone call from my brother, who said a woman had called him from the south Columbus area. She had an alligator that basically outgrew its living area. Jim asked me to get the alligator and transport it to his zoo.

I borrowed my parents' van and started my drive to Columbus. As I drove, I ran different scenarios through my mind. I had had alligator experiences before, but this time I was given no information except the words, 'The alligator outgrew his living area.'

When I finally arrived at the small town south of Columbus, I saw a neighborhood primarily populated by a motorcycle gang. Cycles were parked everywhere—in the yards, on the street, and on the sidewalks. I looked for the address and as I drove down the main street. I collected hostile stares from what seemed to be the entire neighborhood.

Finally, I found the address. As I pulled into the driveway, I slalomed through cycles—and people—lying in the grass and driveway area. I got out and attempted to start a conversation with a man sitting on the front porch. He appeared to be heavily intoxicated, so I walked to the front door, which was wide open, with people going in and out.

As I walked in, the odor of marijuana smoke wrapped

itself around my head. A partially dressed woman approached me with a cigarette in one hand and a bottle of whiskey in the other.

She said she was the owner of the property and she wanted to know if I was a cop. "Yes," I said, and the place went deathly quiet. "Yes," I quickly responded, "I *am* a cop but not locally. I am the guy for the alligator." Meaning, of course, that I was not prepared to notice that these folks were smoking a bale of marijuana, or whether it violated any local ordinances in whatever godforsaken place I found myself.

She smiled and yelled, "The alligator dude!" and gave me a hug as she pulled me to the basement door in the kitchen.

Everyone else followed, although unsteadily. I asked her for some information about the alligator and she said one of her boyfriends brought it back from Florida years ago.

Now, she said, it was close to six feet long. She also said that the gator had lived in the basement for the last three years and that they had thrown chickens down the stairs to feed it.

The gator had escaped a wading pool in the front yard when he was three feet long and scared the neighbors so the sheriff was notified. They responded by trying to kill the gator by beating it in the head with clubs, which, judging from the living room, was the commonly accepted manner of problem-solving in the neighborhood.

Apparently, however, the clubbing did not kill the gator. It only broke his jaw and made him mad. She rescued her gator and ever since, it had lived in the basement. I began to walk slowly down the basement steps with her behind me, hanging on to my shoulders.

As we came closer to the floor, a bellow nearly popped my eardrums, and I saw the gator charging towards us, snapping and hissing. The young lady immediately let go of me and raced back up the stairs where a large crowd of spectators had gathered to watch Gator Dude either catch the gator or die in the basement, the latest in a long line of poor creatures flung down the basement stairs. The crowd appeared to favor the latter.

I growled back and stomped my feet on the lower step. Then I let out my best redneck yell and began to charge back at the gator, startling him. He turned and ran back into an adjacent room.

I slowly approached and found his tail sticking out from under an old bed. I saw that he *was* nearly six feet, but he was also very thick and heavy looking. Growls were coming from under the bed.

I grabbed the tail and pulled him out into the open basement area, and he swung around and tried to snap at me. I swung him by his tail towards the corner of the room and immediately pounced on his back and pinned his mouth to the floor.

I yelled for the woman to bring me the duct tape, which I always carry when I enter a den of marijuana-smoking motorcycle maniacs who have an alligator in their basement. I heard nothing but all the spectators clapping and yelling.

The woman finally heard me yelling, brought the tape, and helped me secure the gator's mouth. When I tried to pick him up, I saw he was much heavier than I thought.

The owner tried to help me by grabbing the thrashing tail.

We then yelled for a couple of the many well-muscled men upstairs but no one would help. As we struggled up the steps, the gator began to thrash around again, hissing and growling. We carried this gator through the living room, scattering the guests as we moved. As we went out the front door, the alligator began to thrash his tail violently, the owner let go, and I fell on top of the gator, trying to both subdue and calm him. My impression was that he was not enamored of motorcyclists.

I yelled at the owner to open the back door of my van, which she hurriedly did. I scooped up the gator, tossed him into the back of the van, and quickly shut the doors. The owner came up and gave me a hug.

"What a rush, man!" she said, breathlessly. I bid the gator's owner a fond farewell, since he was thrashing violently in the back of the van, rocking the vehicle from side to side.

The ride back to Dayton was interesting, as I had to watch constantly to see that the gator did not pull the tape off his mouth and jump up in the front seat with me. When I got back to Dayton, I pulled into my driveway and my boys came out to see him. Neighbors came by to have their picture taken with him. Afterward when I had secured the gator so that he could not move, our family drove off to Kentucky to meet my brother at the Florence Mall parking lot.

When we arrived, we saw Jim and his wife in an open area of the parking lot. I pulled up next to him, we exchanged our hellos, and Jim wanted to put the gator into his vehicle. I warned him that the gator was a fighter and used his tail like a ball bat. Jim laughed and said he had handled all kinds of gators and that this one was no different.

He opened the back doors of the van, grabbed the secured gator, and began to pick him up. Suddenly, the gator began to thrash violently, whipping his tail, striking Jim in the side of the face, knocking off his glasses, and cracking his nose.

Jim said later that this gator was the most aggressive one he had ever been around. What also made this gator different was that his mouth was misshapen by the beating he had received when he was young. His jaw grew at a weird angle, allowing him to be able to stab you with the teeth of his lower jaw—even if you had his mouth pinned shut.

Jim kept him at his zoo for a few years and until he died, he never calmed down. We still think that this gator was the meanest gator we have ever dealt with, a match, no doubt, for the company he kept in his lonely house outside of Columbus.

Crawl Of The Wild

One evening, as my wife and I and some friends were walking into a movie theater, I was paged by the local animal control officers. They had been called to a pet store where neighbors were worried because a horrible smell of decay was coming from the walls and ceilings of the store and traveling into adjacent apartments and businesses.

They called me because in the past they knew that there were cobras and rattlesnakes for sale in the store, as well as two four-foot alligators, along with unknown snakes, scorpions, and tarantulas. I agreed to help and we decided that early the next morning would be soon enough.

The next morning, I arrived to find all the news stations already there, although no one had gone into the pet store. On gaining entry, the officers said they feared that the smell was from the body of the owner, who had possibly died from a venomous snake bite some days before.

What we found was sickening. There were dead snakes in cages, trash cans, and loose. Some of the snakes had decayed into their wooden cages where snake and wood had become one. There were iguanas and geckos loose in the store, as well as snakes and assorted tarantulas and scorpions.

What worried the officers was the four-foot gator in the makeshift pond in the front of the store and his brother loose in the back near some water troughs. I grabbed the one in the front window by the tail and he whipped back and almost bit me in the face.

I flung the gator onto the floor and pinned his mouth down. My son, Adam, taped the gator's mouth shut, and after securing him, we placed the gator in the back of our van.

Then we went to the back of the store where we found the other gator in a water trough, under water. It was dark back near the trough, and the camera man from a local television station turned on his camera light, and I could make out a silhouette in the water, lying on the bottom of the trough.

I had to guess which end was the head and which was the tail. Reaching into the water I quickly grabbed what I thought was the gator's neck area and flung him out of the water and onto the floor. Fortunately, I had guessed correctly.

Adam quickly wrapped duct tape around the gator's mouth, which I had squeezed closed with my hands. We secured the gator and placed him in my van. We caught two alligators, numerous tarantulas, scorpions, iguanas, monitor lizards, assorted small mammals, and turtles, as well as numerous snakes. There were many dead snakes, turtles, lizards, and mammals. It was a sad situation of human neglect.

The gators and snakes were taken to the Kentucky Reptile Zoo where my brother Jim cared for them. The owner, who had skipped town just ahead of drug charges, was cited. Pleading to a lower charge, he will not sell exotic animals again, at least in Dayton, Ohio.

Gator

WHEN SOMEONE GOES INTO A PET STORE and sees a foot-long alligator, they honestly believe that this little critter will make a good pet, especially after the pet store clerk tells them how easy they are to take care of. The clerks fail to tell them that a gator can grow up to six feet long in five or six years, eating rabbits and other expensive food. They outgrow their living facilities quickly and become aggressive and unmanageable.

When this happens, the disenchanted pet owner usually looks for a zoo to take his pet, but no zoo wants the pet. It is a tiny—and might I add, psychotic—minority who wishes to have an expensive-to-feed, aggressive, six-foot alligator.

As a last resort, most gators in this situation are killed or left to starve to death. Some may be released into a river, pond, or lake— to the surprise of people enjoying a nature walk when they suddenly spot a six-foot gator floating past.

Such an incident happened on the Miami River near Dayton when fishermen saw a gator floating by. They notified the police and game wardens, but no one took their story seriously until two young boys, fishing in the same area, brought home a two-foot gator—to their mother's chagrin.

The boys had been fishing in a boat and saw the gator floating in the ice-cold water of an early March morning. They paddled up to the gator, then reached in and picked him up. The gator was nearly frozen, in bad health, and put up no fight.

This discovery started a frenzy of sightseers heading toward that location, where more gators were spotted. A local radio station put up a $100,000 bounty to anyone who could catch or kill a six-foot gator from that part of the river.

Game hunters came from all around the state, as well as Indiana, Michigan, Pennsylvania, and Kentucky. Many of them had high-powered rifles, and local law enforcement officials immediately banned people from the area.

I was notified by a local game official, who asked for my help to stop the madness. I called a good friend, Bill Cacciolfi, a world traveler and explorer who owns New World Expeditions. At New World Expeditions, Bill takes explorers on expeditions to parts of the world where tourists do not go, and he has a great knowledge of tracking critters.

We drove over to the area where the gators were spotted and noticed a small bait shop along the river and two small houses behind the shop. As I walked into the shop, I immediately felt distrust from the shop owners. "More clowns to bother us about those gators," said their suspicious looks.

I introduced Bill and myself and asked him about the fishing on the river. "There ain't no fishing!" he said. "No one goes on the river cause of them gators and the idiots that are hunting them!"

I asked him his opinion about the situation. He was not at all happy about it, he said. His sale of fishing equipment and bait had stopped when the television and radio stations started broadcasting about the gators. He said he had never seen a gator on the river but that his partner, an elderly man with sharp clear eyes and a dry wit who looked to be in his 80s, had seen them.

I explained that we were only here to verify if there were gators, that we had nothing to do with the news media, and that we were not here to hunt them. The two men saw that we were not a threat and the elderly man told Bill and me to follow him and he would show us a place where the gators congregated.

We followed him out to a four-wheel drive vehicle and he said, "Jump in." Bill and I did not understand that we would be taking a ride back into an area of the river that was almost totally wild, with woods and high grass fields surrounding the islands and washed away areas of the river.

This area was close to a water treatment plant where warm water constantly comes out of the plant into that part of the river. The old man told us that he was originally from the southern part of the United States where gators were found and he said he knew a gator when he saw one.

We got out of the truck and approached an island where water had been trapped from the river. Bill saw tracks immediately. The old man said, "See the tail marks? Those are gator tracks. That ain't no turtle."

Bill laughed. He understood that the old man knew what he was talking about. Bill found more tracks and verified that there was more than one gator and that the biggest one would probably be around three feet long. We checked the area but it was so vast that there was no way we could find the gators unless we were lucky, so I notified the officials of what we had found.

Within the next four weeks, they removed three gators, the biggest being about three feet long and the smallest about one and a half feet. Our theory on why there were so many gators at this one site is that people were dumping their pets into the Miami River and since the waters were extremely warm on these islands, the gators traveled down the river, ended up at this spot, and stayed.

Since the gators have been removed, we have checked back over the years and no one has seen or heard rumors of any gators in that area since. I kept one of the captured gators, and we named him Gary. He was a great ambassador for not having gators as pets.

At presentations, I bring Gary up close where I show the kids his massive tail swinging, his size, his teeth. "This," I tell them, "is not a pet. It is a predator." I tell them he is too big and too dangerous. And Gary himself was thrown away because he became so big and so expensive to feed.

The other gators were not as lucky as Gary. They were in poor condition and died. To me they are also ambassadors for not having gators as pets.

Underwater Skirmish Routs Reptile; No Happy Ending

My most exciting gator story is when I was contacted by a city south of Dayton about a six-foot gator that had been seen by numerous witnesses. These witnesses also said that a duck had been found, bitten in half. The pond owner said he usually had ducks, geese, and turtles in his pond but for the last month none had been seen.

The owner said he and a friend were the first to spot the gator, seeing it swim away from the shore, its tail splashing the water, and clearly visible to them. The animal control officials, the local police and I all checked the pond. There were definite signs of a large predator existing in this small pond, which was about a half of an acre in circumference.

I decided I would have to snorkel the pond to see if I could find any turtles or turtle shells cracked open by the predator. Turtles are a good food source for gators and this would definitely prove that a gator was living in this pond.

The nearby ponds were active with ducks, geese, and turtles, but this one wasn't. Witnesses said they saw an alligator, and they seemed honest in their observations. The history of people placing unwanted pets in rivers, lakes, and ponds, and the easy access to this pond by a major road, all added up to a gator present.

I put on my wet suit, mask, and snorkel, and cruised around the pond which was thick around the shore areas with moss and plants. After two hours of searching, I came across a crunched turtle shell near the area where all the witnesses said they had seen a gator.

I immediately noticed stirring on the bottom in the mud, about ten feet below me, causing particles and mud to rise to the surface. I dove down and saw a swirl in the moss area.

"Throw me a rope," I said to my friend, John McCalister, when I came back up. John tied a perfect noose-type loop in the rope and tossed it to me. I dove back down and saw the alligator lying on the bottom in the moss and thinking he was completely camouflaged.

I slowly started to put the noose around its head and he immediately exploded through the thick binding moss towards the shore. The noose set like when a rodeo cowboy lassoes a calf, and I pulled hard, tightening the noose and pulling the gator toward me in the moss.

Immediately, I grabbed the gator by the tail and yanked him through the moss with the rope tied to my wrist. I grabbed the gator by the neck, pushed him deeper into the thick clinging moss, and secured his mouth.

Then came the hard part. I had to try and walk through the moss toward the shore with an angry alligator tied to my arm—the ultimate solution for a bad neighborhood, where you explain to the bullies cowering behind the dumpster that you are merely taking your alligator out for some air.

As I approached the shore, I dropped the head and held the tail as I ripped off my mask and snorkel, throwing them onto the grass. When I tried to pull the gator out of the water, he whipped around and almost grabbed me in his powerful jaws, causing me to spin him in the air and throw him onto the bank.

I held onto the tail and found that the rope was tied around both of us. I approached from behind and pounced on him and pinned his mouth down, yelling for John to get the duct tape.

I was tired. My hands were shaking from the capture, the strenuous exit through the moss, and the near-miss from the gator's jaws. We secured him and notified the property owners, animal control officials, and the police. The owners were happy and relieved that the gator had been captured and their pets and children would be safe. A local television station received a first place national award for their coverage of the capture.

My reward was that now I have great place to go fishing. There was no reward for the gator, however. He had not been properly fed and his feet were damaged, as if he had been kept on a concrete floor. There were holes in his skin, which indicated someone had poked him with a sharp object. He was not living a natural life, and once again that superior creature, man, had done an inferior job with the creatures that had been entrusted to him.

I give numerous lectures to schools, adult groups, and nature centers about exotic animals as pets. I learned from an employee of a local pet store—a national franchise—that they have taken gators out of their stores in this area because so many people, especially children, have complained.

I am very proud of that happening and I feel I have accomplished something. My reward for all the hours I have spent lecturing and taking care of other people's mistakes is that maybe in the near future no one will ever be called out to rescue an exotic animal. That is the goal.

f • o • u • r

A Dragon In The Park And Other Exotic Tales

Elephant Headstand

My first run-in with an elephant happened while I was still working in the emergency room of a local hospital north of Dayton. I saw the ambulance pull up and the crew was racing around removing a patient who appeared not to be breathing.

Hurriedly, they pushed the cot with the patient through the doors to the ER and one of the ambulance crew screamed, "He's been crushed by an elephant!"

I have worked on many unusual patients, but I had never heard the word "elephant" used by any medic crew before. When I examined the patient I did not know what to expect.

The leader of the medics told us how this man was crushed by an elephant. It happened at a circus performing in a small community west of Dayton. The man was the trainer of the circus' Asian elephant.

The elephant had knocked the trainer to the ground and struck him with his trunk and feet. Apparently, the large Asian elephant finally tired of being hooked behind the ears, and struck on the head, legs, and back with poles, which was the trainer's methodology for teaching the elephant tricks. And so the elephant finally struck back.

Then the elephant did a headstand, according to the witnesses, except it was on the trainer's chest, crushing and finally killing him. All the trainer's internal organs were damaged, and his ribs and chest bone were crushed.

Elephants in captivity are not usually treated well, and over the years many have attacked their trainers. People who think that circus animals are not tortured by their trainers live in a fantasy world, especially during this time, the 1970s.

If you have ever seen an Asian elephant in the wild, as I have, you would see a magnificent, regal creature. When you see a circus elephant, you usually see a beaten slave that does what his human master demands, performing to keep from being beaten. Occasionally, one of these elephants rebels, turns on his master, maiming or killing him. People are shocked when this happens, although I wonder why it doesn't happen more often.

The police officer that came to the ER to take the report said the elephant escaped and was loose in the local area, and that no one knew where it was. There was an elephant hunt by all local police departments in the area.

Finally, the elephant was spotted in a large pond, on a farm just outside of town. No one had any idea how to get this jumbo critter out of the pond. A veterinarian I knew was contacted, as well as a naturalist from the local nature center. The police and game wardens wanted to dart the elephant and then pull him out of the pond. The veterinarian and I argued against this method, but we were outvoted and the game warden darted the elephant.

The elephant began to fall asleep, and its head went under the water. We tried in vain to raise its head but to no avail. The elephant drowned. There were no heroes in this story, no happy ending, just a dead trainer and an equally dead elephant. No winners, only losers.

Everyone who witnessed this tragedy will never forget what he or she saw and heard. And I take that as the only positive thing that happened in this fiasco.

Elephant Story With Happy Ending

MY OTHER ELEPHANT STORY has a happy ending, but how I got involved is most unusual. In my city, Oakwood, we have an annual parade that has everything from Clydesdale horses to marching bands. This year they were going to have an elephant from a nearby zoo come up for the parade, and then there would be elephant rides at the local high school stadium.

I was assigned as a patrol officer and was watching the crowds when the dispatcher asked me to come to the station. When I pulled up, I knew something was wrong by the way everyone around the chief appeared upset. I immediately reviewed everything I had done that day, hoping I was not the cause of their turmoil.

As I stepped from my cruiser, the chief asked me where the elephant was. I was stunned. I said I did not know I was on the elephant detail, and certainly I did not know I had lost an elephant.

There were no smiles or laughs from the crowd. They glared at me as the cause of the missing elephant, and the reason why he was not here for the parade. The chief told me to go and find the elephant. Immediately, he said.

I jumped back into my cruiser and raced away from the city building, not having any idea where to look. The elephant was supposedly coming from a zoo south of our city, so I deduced they could be transporting the elephant on Interstate 75. So that is where I started.

I traveled south on Interstate 75, leaving Dayton and keeping my eyes open for anything looking like an elephant, or a vehicle large enough to transport an elephant. About fifteen miles south I saw a large trailer stopped beside the road, behind a truck with its hood open.

I turned around, headed back toward the broken down vehicle, and parked behind it with my emergency lights on. A large man, sweating profusely, approached my vehicle, ready to surrender to the situation.

I rolled down my window and asked him if he was the elephant guy. "I give up," he said. "You can have the elephant. I just want out of here." After I calmed him down and looked over the situation, I decided we needed new transportation immediately, because the elephant was thirsty and becoming agitated. If the animal decided to look elsewhere for water, the trailer was not going to hold her.

I called our dispatcher for a tow service, and it rushed out to assist us. Numerous Good Samaritans stopped to help and brought water, while I stayed in the back of the trailer, keeping the elephant cooled down. When the calvary arrived, the tow truck driver did more than tow the vehicle; he checked the broken truck and got it started.

The elephant was finally delivered and everyone was

happy. We missed the parade, but the elephant was so happy to get out of the hot trailer that she hugged me with her trunk and wiped snot all over my police uniform.

It was the messiest thank you I ever had.

Water Monitor Terrifies Midwestern Folk; Lizard Comes To No Good End

ONE HOT SUMMER DAY, a message left on my answering machine caught my ear and left me questioning the caller's sobriety. The caller was an animal control officer in a city across the border in Indiana, and she said they had a dragon in a pond.

I called, said I could come over that afternoon, and she seemed relieved. The witness reported the dragon was ten feet or longer and looked very much like a dinosaur, scaring the pond owner and the neighbors. She said the head looked long and snakelike, with a tongue like a snake. It was a dull yellow with black markings along the neck.

This description sounded like a water monitor lizard, which is originally from Asia. Pet stores and animal dealers sell these animals when they are about one or two feet long, and if they are fed and cared for properly, they can grow over seven feet long.

These lizards have strong jaws, sharp teeth and claws, and can be extremely aggressive. They look like a smaller version of a Komodo dragon. My theory was that the dragon was a pet and had either escaped or was turned loose.

The officer said there were many hunters trying to kill the dragon with rifles, and that it sounded like a war around the pond. She and the local police feared someone was going to shoot and kill an innocent bystander.

If the dragon was over ten feet, I needed help. I talked some of my fellow public safety officers into volunteering. I felt that if I used guys with whom I fought fire, arrested bad guys, and worked on injured people, then I could not have better protection. If something did go wrong, then I had my own paramedics.

When we left Ohio, I had two assistants energized and ready to hunt dragon in Indiana. When we finally drove through the small downtown area, I made a mental bet with myself that these Midwestern folk had never seen or heard of a water monitor and that a lizard nearly ten feet long would definitely look like a dragon. I could understand their fear of the unknown and see their urgency in wanting the lizard either killed or captured.

We followed the map provided by the animal control officer and arrived at the pond at the arranged time to meet her. There was a large crowd near a tree beside the pond, and I told my partners that this did not look good. The animal control officer came running up to us with a huge smile, exclaiming that one of the hunters had killed the dragon. The whole town had come out to see the monster, and the hunter that killed it was a local hero.

This, I thought, is how the knights must have been applauded when they slayed their dragons. The crowd parted, allowing us to walk up to the dead creature. It was hanging by its neck from a tree branch tied by a rope, as if the animal had been winched.

As I examined it, I immediately saw it was a monitor lizard, close to seven feet long, although with typical poetic license the local newspaper proclaimed it to be twelve feet. The monitor had been shot in the area of its head, probably killing it instantly. The hero said he shot when the dragon raised his head out of the water—looking for something to kill, he thought—so he took careful aim and fired.

What is sad about this story is that this magnificent lizard never should have been there in the first place. When a lizard reaches such a size, it is no longer manageable. It becomes impossible to handle and expensive to feed and care for. This monitor apparently had outgrown his owner's need for him, and it had little chance of survival on its own.

I do not blame the town or the hunter for shooting the monitor. The townspeople did not know better and felt they were protecting their families. I blame the lizard's owner and the pet store or dealer who sold it to him. They slayed the dragon, not the hunter.

It Isn't Easy Being Green

I WAS ONCE CALLED BY A LOCAL animal control officer to a large metropolitan park where witnesses said a huge green lizard was scaring people when it ran across the playground area. It had been doing this every day for a week. The officer said the lizard appeared to be a large green iguana, and they had tried to capture it with snares, nooses, and nets but were unable to get close enough before it scurried up a tree, faster than a squirrel.

When I went to the park, I knew I would need someone else to help me chase the lizard, so that when it went for its favorite tree, I would be waiting out of sight and grab him as he started up the tree.

I recruited a lady in the park who was not afraid of iguanas and had her on call until we spotted the lizard. It did not take long, since the officer had placed strawberries and fruit in a pan near the tree.

I heard a child scream and I signaled to my assistant. Hiding behind the tree near the fruit, I had my assistant chase the lizard. She ran, screaming, toward the iguana, and it bolted right toward my tree. The iguana glided across the ground and up the tree, twisted around to the back of the tree, out of sight of my assistant. I was waiting for him.

I grabbed him firmly around the neck and pulled him from the tree. The iguana was mad and began to whip his tail violently toward my face. I grabbed his back hip area, securing and calming him for a while, which gave my assistant some time to bring me my carrying cage. The iguana was close to five feet long and appeared to be in good health. No one knows how long it was loose, but for a while, this lizard was king of the park.

The iguana was given to a local iguana dealer who cared for it and I went home and nursed my iguana tail whip wounds, which stayed red and sore for a week. Iguanas are popular pets, but people need to know that they can grow big and cause serious damage with their teeth or their tail.

I saw medical photos of a young lady attacked by a large male iguana, and it almost ripped her nose off. She had to undergo plastic surgery to repair the damage. Iguanas also have a reputation for carrying salmonella, and in the Midwest area, over the years, reptiles such as turtles and iguanas have died from it.

Once, at a Komodo exhibit, the wind blew tiny bits of fecal matter onto a railing in front of the exhibit and it became covered with salmonella. When the kids left the exhibit and went off to buy food, they all contacted salmonella, becoming very ill.

Pet owners need to get all the facts about their new pets. They cannot expect pet shop clerks to tell them all the facts because sometimes they don't know the facts, or they don't tell the customers for the sake of a sale.

Lack of knowledge about exotic and wild animals has caused many people to put themselves into dangerous situations. There are many stories in which human stupidity causes a dangerous encounter with animals.

When I was doing shows for a local grocery store, Dorothy Lane Market, I was giving a reptile lecture using alligators, snakes, and turtles. One particular turtle was a jumbo snapping turtle named Peaches. I told my colleague, Matt Heib, to watch Peaches, who was placed in a baby pool full of water while I answered questions from the crowd.

Matt asked me why. I told Matt that if he were not watching the spectators approaching the mammoth turtle, someone would most likely try to pet Peaches and get a serious bite.

"No one could be that stupid," Matt said. "Look at Peaches' size and aggressive temperament, snapping and biting every time someone comes near."

"Just keep an eye out for stupid ones," I said.

I turned around and began to answer questions from the crowd. Matt turned to watch me when an attractive well-dressed woman walked past Matt with a three-year old-girl. She went right up to Peaches' baby pool and was about to put her little girl onto Peaches' back.

Matt immediately stopped her and exclaimed, "What are you doing!"

She said she was going to put her daughter on the turtle's back so she could take a picture. Matt could not believe what he was hearing.

I explained to the lady that this turtle was a large predator and could seriously hurt her daughter. The woman laughed and did not believe us.

As the woman and her daughter walked away, I pointed

her out, using her ignorance as a teaching tool to the rest of the crowd. Except for not allowing the small girl to be fed to the turtle, it was otherwise a perfect visual aid to what I had been teaching during my show.

I have heard of parents placing their children on the backs of wild bison in Yellowstone National Park, their children seriously injured when the bison bucked them off, the parents usually butted or gored in the process.

Infrequent Flyer Boards Plane; Excites Passengers

I HAVE RECEIVED MANY UNUSUAL CALLS in which an animal has caused disturbance enough to shut down a business and cause a panic. One call was to the dispatcher at the City of Oakwood Public Safety Department from a major airline at the Dayton International Airport. The airline was requesting help because one of their planes—traveling from Florida, with a stopover at Atlanta—had a lizard or snake aboard that was terrifying the passengers and crew.

Apparently the airline had called animal control, but since they thought the animal might be venomous, the airline was told to call me. While I was making the drive to the airport, I began thinking of all kinds of venomous reptiles some idiot could have brought onto the plane.

The illegal reptile trade is quite prosperous. The main

way smugglers get the animals to the buyers is to put them on their person, under their clothes or in carry-on bags, and sometimes the animals do escape—usually to the terror of fellow passengers.

When I arrived at the airport, I was met by the airport police and escorted to the now-abandoned airplane where three flight attendants were waiting. The male attendant said he thought the critter was poisonous because of a red flap of skin under its neck. This was making everyone excited and upset.

The two female attendants, airport police, and I slowly began checking the last area in which the animal had been seen. From the description by one of the attendants, the animal sounded like an anole, about six inches long at the longest. Anoles are most commonly called chameleons and live in Florida, which is where the plane originated.

The lizard was discovered under a blanket on the floor, and I caught him. One of the flight attendants said they thought a little boy might have brought the lizard on board. I found the little boy, who immediately denied any knowledge of the lizard.

I asked him if he would want the lizard, because I had no one who did want him. He seemed excited and grabbed his backpack from which he removed a Tupperware bowl with holes poked in the lid. When I asked him why he had a container with holes in it, he said he didn't know, that maybe the lizard crawled into his back pack by mistake.

This story reminded me of when I came home with a puppy when I was little and told my parents that the puppy followed me home even though the puppy had a rope tied around his neck and the other end was in my hand.

The boy on the plane was eight years old and scared of getting into trouble, especially since the plane was grounded until the animal was captured. The lizard was now with his owner and I was leaving through the terminal area when the airline terminal attendant announced over the loud speaker, "Officer Tim Harrison of the Oakwood Public Safety Department and animal wrangler has captured the creature and made the plane safe for everyone."

This brought a tremendous round of applause from the passengers. One passenger even bought me a Coke.

f • i • v • e

Invasion Of The Local Critters

Embroglio In Suburbia; Raccoon Inhabits Home Entertainment Center

IT IS ALWAYS EXCITING TO BE CALLED to capture an exotic animal, and the media is always interested in a loose snake or gator. But my most exciting captures are usually native creatures that have made their way, uninvited, onto people's property. As a public safety officer, I have captured hundreds of raccoons, opossums, squirrels, ground hogs, coyotes, assorted birds and reptiles.

A citizen will call the police department, frantically telling us that a raccoon is in their living room, tearing up the furniture and scaring everyone. Raccoons are unique animals that may grow up to thirty pounds, and they have hands and fingers that can undo and open most any container or cupboard. They are also very aggressive and if cornered can seriously hurt an adult human.

On most calls, I assist the raccoon in escaping the home by opening doors and chasing him out. There are occasions, however, where I have to physically capture the belligerent. Once I was called to a house where a woman and her small children were hearing growling noises coming from behind their home entertainment center.

My fellow officer had already arrived and could also hear the growling. I looked behind the entertainment center, shining my flashlight, and immediately I saw two eyes shining back at me. I knew the intruder was a large raccoon, and he did not want to leave his hiding place.

I told the other officer to open the front door and stand back, keeping the family in the other room. I slowly moved the entertainment center, just enough to reach behind the television and grab the raccoon behind the neck.

This triggered the most awful yowling I have ever heard come out of a raccoon. I secured him by grabbing the tail area with my other hand. I had him by his neck and his back legs, and I rushed towards the open front door, tossing him unceremoniously onto the front lawn. The raccoon ran as fast as his legs could carry him into some trees, plainly unhappy to be evicted from his cozy rent-free spot with the cable hookup.

Raccoons are innovative creatures, getting into people's homes by climbing down a chimney where the fireplace does not have a protective screen. Putting protective screens on the top of chimneys and closing flues when the fireplace is not in use will usually keep out large critters, although small ones can gain entry through the most unnoticed openings.

I have been involved with many calls where a family of raccoons has made its home in a chimney or attic area. Evicting the guests can sometimes be extremely difficult. Once, an officer and I went to the home of an elderly lady who lived alone, and we found that a mother raccoon and three babies had made a home in her chimney.

We contacted a varmint exterminator who said he was going to have to kill them to get them out, and the lady, my fellow officer, and I did not want that to happen. The other officer got a ladder and climbed up on the roof to gain access to the chimney. I went into the house and stationed myself at the fireplace.

The officer on the roof then lowered a chimney chain (used to clean chimneys when there was a chimney fire) which scared the babies. They shot out of the fireplace, growling and snarling. I grabbed them as they came out and placed them into a cage.

We now began waiting for the mother, who would not come out as easily as her babies. All I could hear was her snarling, biting at the chain and fighting this menace to the well-being of her family. So I began to consider a different way to get her down.

I made noises like a baby raccoon in trouble. The other babies started to mimic what I was doing, and I could hear the mother calling back and getting closer to the fireplace entrance. She was near the entrance, but she would only stick her nose down.

I remained still and out of sight. I could see her inching her way out to her babies. As soon as she saw them, she lowered herself down into the fireplace. I lunged and quickly grabbed mom.

Mom raccoon began to fight, rolling around and trying to tear me apart. I kept relatively calm, holding her neck and tail firmly, picking her up, and putting her into the cage with her babies.

The officer on the roof and the owner said that they could only hear the struggle and the bloodcurdling snarls. They thought that surely I was seriously hurt, and they were happy to see that I remained in one piece and that the raccoon family was safely in the cage.

After a hug from the owner, we removed the cage, transported the family to a secluded, well-wooded area, and turned them loose. After making sure that her babies were safe, the mother looked back at us. It was an inscrutable look, which said either that she was thankful we had helped her, or that she wished to take a chunk out of me. I'd like to think it was a look of thanks.

The Amazing Ritual Of The Two-Can Capture

SQUIRRELS ARE CUTE ANIMALS FOUND in almost every community. They are fun to watch until they get into your homes, cause damage, and try to bite the homeowners. One of my most renowned calls was for a squirrel in a woman's basement laundry room. The squirrel was running amok, scaring her and knocking over detergent and boxes as it ran around the little room.

A fellow officer had already arrived and could not get into the room because of the squirrel's aggressive defensive maneuvers. I told the owner and the officer to step away from the door, and I looked around for something in which to catch this varmint. All I could find were two coffee cans.

I grabbed the two cans, opened the door and quickly rushed in, not giving the squirrel a chance to run out and get into the other area of the house. The squirrel chirped at me, gnashing his teeth in an attempt to try and intimidate me.

When he became aware that I was not going to run away, he began to run in a circle around the shelves and the washing machine and dryer, using the same path every time he circled. I timed the squirrel so that when it hit the washing machine, I slammed the coffee cans together, capturing the squirrel in the two cans. It was a maneuver that amazed me as much as it did the squirrel.

None of us—the officer, the owner, the squirrel nor I—could believe what had just happened. I left the house and released the squirrel near a wooded area. The legend of the capture, even now, occasionally makes the rounds of the department where I am known as the fastest can in the Midwest.

Party-Crasher Menaces Proceedings

Opossums are usually nocturnal creatures that have a nasty set of teeth and can be very aggressive. One of my most favorite opossum calls came on a warm summer evening when one had the temerity to interrupt a formal party at an outside pool.

This full-grown opossum just wandered up to the partygoers, scaring them so badly they all ran into the house and called the police. When I got there, I saw many faces looking out of the windows and doors of the residence and a frantic man pointing toward a food-laden table.

I walked up and the opossum immediately started to growl, putting on a little show to scare me off. When I walked closer and it saw I was not intimidated, he showed his teeth, going into a defensive mode and backing his rear end up against a cooler.

I faked it by acting as though I were reaching for its neck. The opossum swung around and tried to bite my left hand, exposing its tail, which I grabbed with my right hand, lifting the critter up. Then I carried the opossum over to a cage I had brought with me.

The partygoers all came out to see the beast that had frightened them. They began to slap me on the back and praise my animal capturing skills. I smiled as each person praised me and walked over to the food table and picked up some food for my opossum friend in the cage, which was, after all, why he had come.

After waving good-bye to the crowd, I relocated the opossum—and his food—to a secluded wooded area. It still amazes me why people fear these animals, but I have understood that their fear is from lack of knowledge about their animal neighbors. Most people have no idea who or what lives in their own yard. How boring not to have this information. How much more fun life would be, knowing *all* of one's neighbors.

Bat Man Gears Up For Mortal Combat

THE PREDOMINANCE OF CALLS over my eccentric career have been calls about the brown bat. The brown bat is a beautifully constructed creature, a flying mammal. Flying squirrels do not actually fly. They glide, which, I suppose, makes them a gliding mammal.

The bat is very important to our environment as it is a magnificent eater of mosquitos, dining on hundreds at one meal. These animals are an important link to the chain of life, but most people consider them flying vermin and carriers of deadly rabies.

I always remember the scene from the movie, *The Great Outdoors*, when John Candy and Dan Akyrod discover a bat loose in their cottage and to combat the vast menace, put on combat gear. Their reaction is the reaction of the average person confronted with a bat.

I have been on hundreds of bat calls, but my most memorable call was when the editor of a local newspaper wanted to see Batman work his magic. Catching a bat is not easy, but I had a reputation of catching bats with a shoebox.

My technique was to cause the bat to fly, then I snagged it in the air with the box and slammed the lid back on before it could get away. This procedure worked about ninety percent of the time, although there are some bats that don't fall for that.

When I arrived at the editor's home, we walked up to the bedroom where the bat was last seen, and as we entered the room we could see the bat on a window shade.

I asked for a shoebox. My reputation had preceded me—the owner already had one for me. I approached the bat and quickly slammed the box over it. The window shade bowed inwards toward the window giving the bat a space to escape between the box and the shade.

The bat began to fly around the room and the editor and I saw that this was one large brown bat—almost a foot of wing span. I yelled at him to block the doorway so that the bat could not escape into the rest of the house. He was reluctant but waved his arms and managed to keep the bat from escaping.

I tried my shoebox snare in the air, but this bat was not going to let me catch him so easily. I chased the bat until both of us were tired. When he came to rest on the floor, I slowly put the shoebox over him. I slid the lid under the box and he was secured. The bat was caught but the owner was a bit disappointed because he had wanted to watch the famous midair capture.

I laughed and told him maybe we could do this again sometime, exiting with my shoebox and great respect for this elusive bat. The bat was released in a wooded area and the editor wrote a nice article in his newspaper about our adventure.

Bats turn up in the most outlandish places. Once, on street patrol, I got a call that a bat was in a swimming pool at one of our city parks. As I responded, I searched my memory but as far as I knew bats did not swim.

Everyone was far away from the pool. It was a wading pool for small kids and there was a medium-sized brown bat floating and trying to do his best to swim. (He appeared to be doing the batstroke.) I reached into the water and cupped him with my hands, at which time he began to bite and try to escape although he had no energy left to do much of anything.

The bat was brought back to the station where I placed him into a box to recuperate from his ordeal. The bat became stronger and I released him later that evening. How he got into the pool is only a guess, but I guessed it must have been out the evening before and found the ledge around the pool to be a nice place to sleep off the day. When the kids jumped in, they woke up the bat, which probably panicked and fell directly into the pool. Now I have a new title: Pool Batguard.

Close Encounters With The Coyote Kind

When people think of coyotes they think of the west and remote areas, but coyotes have found new homes in Midwestern suburban areas. There are numerous reports of coyotes spotted in alleys and around homes, where they eat trash, or pet food left out for the family pet—or maybe the family pet itself.

Coyotes can be nuisances, and they have attacked children and adults across the county. In Yellowstone National Park, coyotes having taken food right out of campers' hands.

Coyotes have become bold, even chasing people from their picnics. In my area, we have had numerous calls of coyotes lurking around playgrounds and running in plain view across major streets. They are animals that adapt well and can survive any circumstance or situation. Some scientists have said that if we had a nuclear war coyotes and cockroaches would inherit the earth.

The antidote for keeping coyotes out of your neighborhood is to take away their food sources. When there is nothing to eat, they will leave. People have to seal up their trash in heavy trashcans or receptacles so there is no odor and it is impossible for any critter to gain access.

Do not feed your pets outside, and never leave food out for them. This attracts critters that are not friendly to your pet. Coyotes love to eat cats. Many city cats are de-clawed and wear bells to scare away birds. To the coyote, this is the sound of a dinnerbell.

Never feed wild animals at your home. Animals associate food with people and will approach other people expecting food. When they have none, the animals become aggressive. In Florida, a large alligator killed a three-year-old girl after she was left alone by her parents to feed the alligator chicken wings. When the wings ran out, the alligator was still hungry and attacked the child, relating her to food. The parents were sitting on a porch drinking a few brews about two hundred yards away.

Wardens shot the alligator instead of the parents, who received no reprimand for their ignorance except the loss of their daughter. This story is sad but explains how wild animals think and people do not.

People ask why coyotes have suddenly appeared in suburban Ohio. I answer that they have always been here, out in the wooded areas that have suddenly become acres of plat homes. This causes the coyote to adapt, moving to where the food is and becoming more acclimated to human beings.

Most coyotes would never see a human, but humans have invaded their turf, bringing gourmet trash, which is much more appealing to a coyote's palate than a rabbit. The coyotes are forced out of their own areas by new homeowners and dogs, and now they are—like most everyone else in America—adapting to suburban life where there is no predator to kill them except man.

One of my close encounters with a coyote occurred on a cold blizzardy night in February. I was on street patrol when my dispatcher asked me to investigate a report from a citizen who said he was looking out his living room window and watching a wolf eating trash from a trash bag in the middle of his street.

I immediately thought that the man was watching a husky or a wolf hybrid type dog, but the street crews, out plowing snow, came across the radio and said that I should go back to the police station and get a silver bullet. They, too, saw a wolf in the middle of this suburban street.

I rushed over to the area, and when I arrived, the neighbors wanted me to shoot the wolf. I walked up to the wolf and saw that it was not a wolf but a large coyote, having supper with someone's unsecured trash.

As I approached, the animal demonstrated poor table manners. It showed its teeth and snarled. I quickly moved him and kicked him hard in the rearend, sending him flying into a snow bank a few feet away.

Having established myself as the alpha male, the coyote ran toward a wooded bike path area that was behind this row of homes. I followed him, shining my flashlight in his face and yelling, hoping to scare him off permanently.

He put his tail between his legs and ran as fast as he could away from the homes. For the rest of the winter, we had no more calls about wolves or coyotes in our community. In the community that borders our city, calls about coyotes continued through the summer, but our coyote dining room remained deserted during that same time frame. The big male and I seemed to have come to an understanding that one winter night.

Rowdy Raccoon Rumble Raises Revolt; No Rapture For Raptor

WHEN I FIRST STARTED to respond to animal rescue calls back in the middle 1970s, there were no animal rehabilitation specialists. A captured bird of prey was taken care of by me until it could be released or placed in a nature park or zoo. Over the years, I have rescued and rehabilitated close to one hundred raptors including sparrow hawks, kestrels, red tail hawks, and all kinds of owls.

My first raptor call was from a local sheriff's office. The sheriff had called the veterinarian for whom I was working at the time, and said an eagle had been shot and going crazy in a suburban yard. I took along, as tools, a large dog cage and a towel.

There was a large crowd in the yard area, excited and pointing to an object in the grass. As I made my way through the crowd, I saw the injured raptor and recognized that it was a red tailed hawk, not an eagle. It was scared and attacking everyone who came close to it.

I made the crowd disperse so I could try to calm the hawk. After the people left the yard, I sat down near the hawk and started talking soothingly to her. I slowly inched my way closer until I was sitting next to her. I could see that she had massive trauma to her left wing. She was covered with blood, and the wing was almost totally missing.

If I was right, the hawk would slowly weaken from the blood loss and I could humanely place her in the cage for transport. When she grew lethargic, I slowly picked her up in my towel and placed her in the carrying cage, quickly transporting her to the veterinarian's office.

The veterinarian did surgery on the wing to remove shotgun pellets, closed up the wound, and stopped the bleeding. She was given antibiotics, and now it was up to her. Days went by, she grew stronger, and she began to eat mice that I gave her. She let me place her on my gloved hand—I used an X-ray glove at the time—and she became relaxed. I think she actually liked being with us.

When healed, she was placed in a local nature center and used as a demonstration during lectures. She became an ambassador for her fellow hawks, where children and adults could see the beauty and magnificence of these birds.

In the 1970s, many farmers and hunters believed that hawks killed their chickens and livestock and that they should be exterminated. The red tailed hawk was on the endangered species list and its future looked grim. The overall feeling about hawks was that the only good hawk was a dead hawk. Hawks were shot, poisoned, and their young destroyed.

It is sad to say that red tailed hawks (now doing very well and can be seen along interstates hunting mice and small mammals) were almost eliminated from our state. I believe that our work and lectures educating the public with our injured hawks actually changed the minds of many farmers and hunters, enabling them to regain their population.

At that time, there were few animal organizations and even fewer people who cared about what happened to our animal neighbors. It was exciting to see humans making major decisions to save whales and dolphins, and to start protecting eagles and raptors. It was the turning point in the survival of our natural world.

Sometimes my home would be a small animal hospital, with many injured animals being taken care of at the same time. My mother understood, but I don't think she was happy with my patients. Many times, one critter would try to get another critter.

Once, I was rehabilitating a kestrel after it had mistakenly flown into a large glass window and knocked itself out. The hawk was almost ready to go back into the wild and I had made plans to release him. I allowed the little hawk to fly around in our basement. One day, when the little fellow was loose I forgot that I had three young raccoons out and playing in the same basement.

Suddenly, there was a terrible screeching noise. I ran down the steps to the basement where I saw the raccoons climbing up the curtains of the glass basement door trying to get at the kestrel, which was perched and screeching on top of the curtain rod.

The raccoons were tearing down the curtains with their body weight, and the kestrel began flying around the basement landing on different perches, causing the raccoons to race over and try to get at the kestrel.

The raccoon is an incredibly curious creature.

Introduce anything in its terrain, and the raccoon will soon be investigating. The kestrel, too small to harm a raccoon, was merely a flying toy, something to be investigated. The entire basement was a disaster area. After a mad chase, which caused more damage, I finally caught all the raccoons.

I placed the raccoons back into their cage and tried to calm the kestrel. The poor hawk's heart was beating faster than I could count. The hawk finally calmed down and I placed her into her cage upstairs. Calming down my mom was more difficult, and I had no cage large enough to contain her.

A few days later, the kestrel was released into the wild. The raccoons grew up strong and I was able to release them a few months later—leaving only my mom, who was still a bit ruffled but had gotten more docile.

Great horned owls are magnificent raptors, which, oddly, are part homing pigeon. I once captured a great horned owl wounded by a gunshot and found in a yard. The bullet wound was not life-threatening, and the owl healed beautifully.

I took the owl out in our back yard to strengthen its wings. The owl would fly and glide around in the yard but always came back to my gloved hand. This was the first great horned owl that ever did this. I became close to this owl and spent a lot of time with him.

When the time finally came for the owl to be released, I took it to a nature center many miles from my home. After I transported him well out into the forest, I released him and he flew high into a large tree. I walked back to my vehicle and drove home.

As I pulled into my driveway, I saw a silhouette on the roof of my house. To my amazement, it was the great horned owl, sitting on my roof waiting for me to get home. When I got out of my vehicle, the owl flew down and landed on my arm. I could not believe that this owl found his way home.

I kept him overnight, then drove to a nature park an hour away and released him. I stayed for an hour or so and watched him. Finally, he flew out of my sight, and I quickly went to my car and left. I almost expected to see the silhouette on the roof again when I got home. He had finally realized that it was time to be wild again. This owl was not the average owl, and I will always treasure the time I had with him.

Great horned owls are very important to the chain of life. They are great eaters of mice and small mammals, in addition to being part of our mythology. The great horned owl, for instance, is the only owl in North America that has killed a human. This doesn't mean that the great horned owl hunts or attacks humans. The death for which it was responsible was a bizarre accident.

Great horned owls love to eat skunks, and a settler had placed a skunkskin hat on his head and was walking through the forest. The owl saw the hat, and thinking he had spotted a skunk, flew down and struck the hat with such force he fractured the man's skull.

When I first read this story I decided never to wear a skunk on my head, and if I did, I should never wear one in or around a wooded area. It was also a cautionary tale about poor taste in headgear, and about an unfortunate skunk who suffered two mortal attacks. I gained a new respect for the power of the great horned owl, the tiger of the skies.

Never Cry Groundhog

My most dangerous encounter with one of our furry, feathered, or scaled friends happened when I was dispatched to a home whose occupants found a large groundhog in one of their window well areas. Groundhogs, having rodent-like teeth, can be extremely vicious and aggressive. The groundhog has the same temperament as a badger or wolverine, and this big fellow was down in a basement window well with no way to escape.

Even as I approached the window well, I could hear him growling and snarling, advertising his unhappiness. The owner had small children and they were afraid to play out in the yard. When I leaned over to check out the groundhog's dilemma, he backed up into the corner and then jumped at me, trying to bite me.

In my mind, I went through every possible way to assist this groundhog, but each way ended with me in the hospital for stitches and reconstructive surgery. Finally, I decided to jump into the well and step on the groundhog, pinning his head down. Then I would grab him and fling him out of the window well. I have actually done this before with young groundhogs and lived to tell about it.

I jumped into the window well. As soon as my feet hit the well bottom, the groundhog charged me and attacked my legs. I leaped into the air, narrowly missing having my legs ripped apart. The groundhog's teeth ripped my right pant leg, as if to announce, "So much for that plan."

This large groundhog was not going to allow me to out-muscle him, so I now had to out-think him. I saw a wide-bladed snow shovel, grabbed it, and jumped back into the window well. The groundhog immediately attacked. I used the shovel blade to block his attacks—like a hockey goalie blocking pucks.

After I tired down the groundhog, my next step was to try and put the shovel blade under him and lift it up level with the ground so he could escape. The window well was about five feet deep, and I would have to raise this beast up to eye level to pitch him out, an idea which left me less than excited.

I slowly slid the blade under the groundhog, which was still snarling and hissing but too tired to fight, and I began to lift up the shovel. When the shovel blade—with the groundhog on it—reached the level of my face, the beast took a swipe at me with its teeth, causing me to toss him over my head.

I looked up to see the groundhog airborne, but with my head tilted back, it appeared as if the groundhog was going to land directly on my head. I put the shovel blade over my head, waiting for him to land on me but I felt nothing.

Slowly, I looked out from under the shovel blade to see the groundhog running off across the yard and into the woods as fast as his stubby legs would carry him. My appreciative audience of homeowner and children applauded from the safety of a side window.

I was never so happy to be finished with a call. Of all the animal calls that I have been on, this was the closest I ever came to being badly bitten. I always enjoyed watching an animal returning to the wild, feeling blessed to be able to assist and rescue these creatures. This time, I felt doubly blessed, once for the rescue of the groundhog, and again for my own.

s • i • x

Who's Afraid Of The Big Bad Wolf?

The Misunderstood Nature Of The Wolf

WOLVES WERE EXTERMINATED from Yellowstone National Park because of people who misunderstood them and feared them. There is a mythology about the wolf rivaling that of the snake, in which wolves, devil creatures, sneak into the homes of humans and capture sleeping children from their beds. And so they were exterminated, with the park rangers, themselves, helping to eradicate them.

Over the years, wolves have returned. Recently, more wolves have been found in local suburban homes than in the wild. When I say wolves, I am also including wolf hybrids. A wolf hybrid can be a mixture of grey wolf and German shepherd, husky, or malamute dogs. I do not include collies and other mixtures because they do not directly carry the wolf traits and usually do not resemble wolves.

The wolf hybrid can be a problem when grey wolf traits are bred into a German shepherd, creating an animal that has the aggressive characteristics of the German shepherd and the cunning and strength of a wolf. This kind of animal has been the major reason for wolf hybrid attacks on humans.

There have been no reports of a wild free-roaming wolf ever attacking a human in the history of North America. People do not understand that a wolf hybrid needs to be treated differently than a domestic dog. Wolf hybrids will tear a couch apart if they smell one single cheese puff under a cushion (although, to be fair to the wolf, I have seen my teenage sons do the same thing). The wolf in the animal will dig and tear, searching out odors or food.

Most owners do not know this pertinent bit of information when purchasing a pup blindly rushing into the transaction, thinking only of the ego rush of owning a wolf. I have been called many times to a location where a wolf is loose in a neighborhood only to find a wolf hybrid at large. After I capture the animal, the owner usually never comes to claim it. It is usually placed in a humane society kennel and eventually destroyed, all because someone had to have a wolf-type dog as a pet.

While the majority of wolf calls I respond to involve wolf hybrids, I have responded to calls when the animal turns out to be an honest-to-goodness wolf. I have raised pure grey wolves from pups up to adulthood and received them as adults and worked with them successfully. At one point, I had a facility that was used strictly as a maternity ward. Nearby zoos and private dealers brought their pregnant wolves to my facility and left them with me until they had their pups.

I also had pure wolves of my own that I used for school and nature center lectures. One such wolf was Honey, a wolf I rescued from euthanasia. Honey was in a zoo where she would not allow anyone to come near her, attacking her handlers, causing the zoo administration to want to get rid of her.

A friend who worked at the zoo called me and said I needed to come right away and save this eight-month-old female wolf. My assistant and I grabbed a large dog traveling cage and headed out. The experts wanted to dart her with tranquilizers, but I have seen many animals react badly to the medication and die. Then they wanted to try and noose her, which would certainly have injured her, causing her to become more combative and dangerous.

With the experts, my friend, and my assistant watching, I entered the area where the wolf was—a large room-type cage with a kennel exit door to the outside and a gate and fencing on the inside. The exit to the outside was closed so the wolf was contained in this twenty foot by twenty foot room.

When I first entered the cage, the wolf raced to the farthest area of the cage and began to show her teeth. I walked over to the middle of the area and sat down in the straw facing the wolf. Immediately, she charged at me, stopping a foot away, snarling and snapping her teeth.

I knew she wouldn't hurt me because she was a young female, and if I became the alpha male to her, she would become submissive. This invariably happens with a young female. If the wolf had been an alpha male himself, I wouldn't have been in the place.

She went back to the farthest part of the cage and began to pace back and forth. Every few minutes, she charged me, but she never came close enough to hurt me or for me to grab her.

This went on for about fifteen minutes. Finally, when she charged me again, I jumped up and charged at her—

causing her to go into a passive, almost shocked, state, rolling over on her back and peeing on herself.

I was now the alpha male, and I put my face next to her and growled, showing my teeth. She became totally passive, to the point where I could pick her up and place her in the traveling cage without incident.

The experts could not believe their eyes and tried to rationalize what they just saw. One decided it was a freak situation and could never happen again. Another said that if I had approached a male he would have torn me up.

I now had a female wolf, which needed some tender care. On the ride home, my assistant, in a state of shock herself, said continually, "I can't believe what I saw!"

I explained to her that each animal was treated differently and that this was not the norm. I go by my feelings—sixth sense, if you will—about the animals with which I am dealing. I read this wolf as an opportunist that would charge and scare people as long as they would allow her. She was feeding off their fears.

I have done the alpha male act before and it has worked even with males. Most wolves need an alpha in charge and I go by the philosophy that when I am sitting down, I look passive and easy to approach. Sometimes this works, getting an animal close enough to collar and walk out.

If they continue to act aggressive I wait my turn and catch them totally by surprise, jumping up and scaring them, turning myself into an alpha male, which causes the wolf to become passive.

On the way home, a three hour ride, the wolf stayed on her back, lying perfectly still in a passive position. When we got back to my facilities, I removed her from the carrying cage and put her in a cage with an indoor area and an outdoor run so she could go in and out when she wished.

The entire time she stayed totally limp and passive, almost as if she were under anesthetic. It was three days before she would walk outside and two weeks before she would not roll over every time she saw me.

Within three months of intensive adaptation therapy, spending every extra minute I had with her, she began to respond to me. She would come up and allow me to rub her and walk her on a leash. Within six months I was using her to give programs to naturalists and adult students. In one year she was going to elementary schools and kindergarten classes, enjoying every person she met, licking each and howling for all the kids. She was given a name based on her disposition. We called her "Honey."

Honey was a great ambassador for wolves in the wild, helping people to understand the difficulty in having them as pets. Even though she was sweet as honey, I would put a small piece of food in a purse, place it on the floor in front of a lecture group, and begin to tell them about the destructive capabilities of a wolf.

Then I would bring in Honey and watch the faces of the group as she tore the purse apart, looking for the little piece of food. I explained that this could happen to your couch when you accidently drop a snack between the cushions and your wolf sniffed it out. I wanted them to understand what a high maintenance animal they are and how they do not make good house pets. If you love wolves, I tell these groups, then leave them in the wild. If you want to see wolves, go to a zoo or an animal park. Wolves should not be tied up on ropes or chains in backyards or held prisoner in cages, just so someone can say that they own a wolf. Their nature precludes them from being owned.

The Fear Biter

I HAVE BEEN INVOLVED with nearly a hundred wolves over the years, and I have never had a situation where I feared for my life—except once. That remarkable exception involved an extremely shy, pure black, pregnant female wolf. She had torn down a fence to try to get to her owner because he was trying to remove her cubs.

This time, the owner asked if I could keep her until she had her pups. I had another pregnant wolf staying with me, as well as an arctic wolf and my own three wolves, Lupus, Honey, and Whitney.

I accepted the challenge and when she arrived I saw that she was a "fear biter," which means she would bite only out of fear for herself or her pups. Everything went well and she had her pups. She allowed me to pick up her pups and showed no signs of aggression.

After six weeks, the owner wanted to remove the pups, and I told him I would take them out and have them ready when he arrived. When I approached the cage, the mother wolf stood up, came over to the gate, and allowed me to rub her neck. I opened the run door and let her into the outdoor run.

Once she was outside, I lowered the door to separate her from her pups then entered the cage area and began to remove them. All the pups seemed healthy, except for a whiney little male that ran away every time I tried to pick him up. Every time the whiney pup let out a yelp, the mother tried to rip open the door separating her from her pups.

I had all the pups out except this one. When I finally cornered him, he began to howl and fight as though I were trying to kill him. The mother's protective instinct kicked in. She shattered the wooden door, charging at me, snarling and growling with a look in her eyes that chilled me.

The indoor cage was only five feet tall and ten feet by ten feet in width, so I had to bend over and crawl into the cage to get the pups. I was on all fours when the mom attacked. She lunged for my face, and my natural reaction was to punch her in the mouth, which caused my fist to strike her teeth.

Her canine entered my right hand between my second and third finger, deeply puncturing my hand, dislocating my knuckle, and causing the tooth to be lodged in my hand. I reached up and grabbed her neck, and she began to roll over in a submissive position.

What saved my hand was that she did not bite down, which would have forced the other lower canine through my hand, possibly ripping my hand apart. She seemed to understand that I was hurt and quickly became submissive.

It was my fault my hand was in her mouth, and it was

dumb for me to ever punch a wolf. The situation developed so fast that I could only react. When the mom was lying still, I slowly pulled my hand out of her mouth, suddenly aware that the whole canine had entered my hand.

It was as though I had been stabbed. Blood began to spray out of the gaping wound. I picked up the pup and left the cage, heading to the bathtub where I had to wash the pup. I did not want to look at my hand but with all the blood pumping out, I thought I should. The wound was about an inch and a half long, gaping open. When I looked down, I could see my knuckle.

I knew I needed to have the wound stitched, and maybe some work done on my knuckle. After I wrapped my hand with gauze and bandages and cleaned the pups, I gave them to the owner and had my wife take me to a surgeon friend.

When I came home, I went out to try to make up with the mama wolf. I entered the building and she immediately went into a passive position and peed on herself. I opened the gate, crawled in with her, and laid down beside her.

I talked softly to her and slowly stroked her face, head, and neck area. She eventually came around, stood up and licked my face, and I realized she was accepting my apology. The owner came back a few weeks later and picked her up. I felt I had learned a valuable lesson about the strong protective instinct a mother has for her pups.

After this episode, I decided I would not take pups away from their mothers. I learned to respect the mother's need to keep her pups close. There is, especially with wolves, a strong maternal bond and we as humans need to understand it.

I have been told that there are no orphans among wild wolves, that another pack will come along and take care of young pups if they find them. We as people could learn a lot from the wolf.

I still receive calls from local police and animal control personnel about wolves loose in suburban areas, scaring people. To this day, the big bad wolf syndrome is very much alive. Even though the public is better educated, people react negatively when they see a hundred-plus pound wolf running through their back yard.

I have caught many of these wolves by simply clearing the area of people, earning the animal's trust, then placing a leash and choker on it and leading it out to my vehicle. There are times when a wolf hybrid, especially one mixed with a German shepherd, will attack people out of fear. Or perhaps it was trained to attack by its owner. These wolves are usually shot and killed before I ever get the chance to respond. A vicious dog, no matter what its mixture, is a dangerous animal. The domestic dog kills more people each year than snakes, sharks, or bees.

When a game warden, police officer, or animal control officer responds to a call about a loose wolf, they must first think of the public safety. Wolf owners should be aware that if their pet gets loose, the chances are very good the wolf will be shot. This is a good reason not to have a wolf as a household pet. If you love wolves, why subject the animal to a death sentence. Who is afraid of the big bad wolf? When its running loose in the neighborhood, everyone is.

Overzealous Exterminator Damages Premises

THE BEST STORY OF HOW destructive a wolf can be to a home is an incident involving a wolf that, according to its owner, "has gone crazy and torn down my living room wall."

When I met the "crazy" wolf, he seemed normal to me. After looking at the damage, I knew something must have triggered his destructive behavior. Wolves have no inherent desire to renovate houses. As the owner and I talked, a small cockroach ran across the floor of the living room.

Immediately, the wolf jumped up and chased the bug toward the baseboard. When the bug scurried under the baseboard, the wolf began to rip at the baseboard with his paws, trying to get to the bug—exactly as a wild wolf would do. The wolf did not go crazy. He was merely looking for a snack. In other words, he was behaving quite naturally.

seven

My Most Dangerous Encounters

Hair-Raising Flashbacks

WHILE TRAVELING THE WORLD WITH international wildlife television crews, authors of adventure books, wildlife researchers, and fellow adventurers, I have had many situations where I found myself in harm's way. When I give lectures, in Dayton, Ohio, or Capetown, South Africa, the most frequently asked question is: "What was your closest encounter?"

A close encounter is any incident that raises the hair on the back of my neck and gives me a massive adrenaline surge. After the encounter, I react as if I were in a mild state of shock, my hands shaking and—rare for me—difficulty talking. I usually have to sit down and think the situation over and over again to get my brain to accept what has just happened.

There have been many close encounters, but there are six that I consider the most dangerous, ones which, even today, still give me flashbacks.

Six

WHEN I SAY CLOSE ENCOUNTER, I mean to say that usually I am the victim of an encounter in which the grace of God rescues me from my own stupidity. My sixth most dangerous encounter happened on an expedition to Big Cypress National Park in Florida.

There, I met my good friend and fellow adventurer, Keith Gad. Keith is a National Park Ranger who works around the country, specifically at Yellowstone National Park and now Big Cypress National Park. Keith helped me produce the television shows, *Grey Wolf Recovery Program*, *Yellowstone National Park* and *Bears of Yellowstone*. Keith said there was a small lake near his living quarters that had over ten alligators in or near it, one of which he estimated to be over ten feet. He thought we could get some interesting video to use for my school lectures. Little did I know that I would be meeting a wild gator in a most surprising way.

Keith said that no one would snorkel with him in the lake, and to his knowledge no one had *ever* snorkeled in the lake. I geared up in a wet suit, weight belt, mask, snorkel, fins, and underwater video camera, and slipped into the water, trying not to scare the animals already checking me out. I was able to snorkel within inches of a five-foot alligator and record the experience. I was quite excited.

Keith was on the shore helping spot gators for me, walking along the bank, following me as I snorkeled along the shoreline. The shoreline consisted of hills that dropped straight down to the water's edge at about a 70° angle, which caused Keith to lose sight of me for short periods of time.

There was a sharp drop-off, making it extremely difficult to stand on the rocky ledges and adjust my gear or to get back up on the shore. As I was snorkeling around a large rock outcrop there was a hill area hidden from Keith as he approached on land. When I rounded the ledge, Keith was just reaching the hill. Thinking all the gators were on the other side of the lake, he was making no attempt to be quiet.

I was filming some fish ahead of me. I was underwater except for the tip of my snorkel when I was struck on the back of my head by a hard, large object that almost knocked me out. I was dazed but suddenly realized a five to six foot alligator had landed on my back and was striking me with his tail, his jaws within inches of my face.

I still had the video camera rolling and I flipped my body around to elbow the gator across the animal's head. He rolled off and was now beside me, still smacking me hard with his tail. Keith raced up the hill and saw our stare-down. I began to yell through my snorkel and push the camera at the gator's face.

The gator decided not to grab me in its jaws and turned quickly, shooting off into the deep, dark lake. I was still stunned from the hit on the back of my head, and I was yelling through my snorkel for Keith to help me get to the shoreline.

He ran down to the water's edge and assisted me as I kept yelling, "Did you see that?!" I could not believe what just happened to me. I looked at Keith and I realized I had caught the encounter on tape. When we got back to Keith's living quarters, we plugged the camera into the television and watched my gator close encounter with the other Rangers.

Apparently this gator had been sitting on the lakeside sunning himself, not a care in the world, when he heard a noise behind him—Keith. Thinking it was going to be safe in the water, the gator raced away from Keith, down the hill, and jumped into the water like he had done all his life without incident. The alligators at this lake had never seen any humans in the water before, but when this unfortunate gator hit the water, he found a six-foot human trespasser. What a surprise for this prehistoric reptile. He must have been as much in shock as I was. The gator could have bitten me but chose not to, only knocking me away with his powerful tail.

I call this occasion, "The Gator Who Went for a Piggyback Ride," and he lives both on video and at Big Cypress National Park.

Five

THE FIFTH MOST memorable encounter happened in Asia on the border of India and Nepal, at a game reserve called Gaida camp. We had just arrived and were taking our bags to our tents when a group of naturalists from Germany began screaming and running towards a wooded area near the tents.

Running separately from the group, I saw a young German couple headed towards some elephant grass that was burned down by a recent fire. As I approached the couple, the young woman screamed in German. I turned, just in time to feel an object swing by my hip

To my considerable surprise, it was a large Asian black bear, upset and lashing out at anything in his way. I had actually run right into him, almost getting ripped apart by the bear's enormous claws. The bear kept running into the low elephant grass. I snapped a picture of him as he turned to look at the group of humans who were chasing him.

Finally, the bear disappeared into the woods and I understood how stupid I was for chasing a wild animal and not paying attention to my surroundings. I was being a tourist.

Four

My fourth most dangerous encounter occurred when I was in Gansbaii, South Africa, filming a show on great white sharks. Off Gansbaii are two islands, Geyser Rock and Dyer Island. One has fur seals, the other has penguins. In between is Shark Alley, a channel of water where great white sharks cruise, looking for prey that is not paying attention—the slow and the weak. Caught as it is between the two islands, it is a shark buffet.

I was on the boat with the Great White Shark Research Institute. We started our day off by chumming the channel and putting chunks of fish on long lines to seduce the sharks up to the boat. It did not take long before a great white shark appeared and started following the bait to the boat.

I had a small video camera in my hands and I began to record the shark as it was chasing the bait towards the boat. As the shark approached the boat's stern, I was looking through the viewfinder. Forgetting how close I was to the side of the boat, I stumbled and fell over the side—still filming—and my camera struck the shark on the nose as he was ripping into the bait.

The shark's teeth missed my hands and face by inches. As I kept filming, a crew member grabbed my belt and pants and pulled me back up on the boat. The shark kept biting and tearing the bait apart. I was so close to his mouth I could smell what he had for dinner, and it smelled like a penguin. This footage was used on my show about great white sharks. Every time I see the video I shudder.

Three

MY THIRD MOST frightening situation involving wildlife occurred when I was at a game reserve called Sukula Phante in western Nepal, near India. My partner, Tim Werbrich, and our guide, Santoosh Bismet, were walking on a dirt path headed towards a machua, a Nepalese tree house. We noticed the two other members of our team were already there, waving their arms wildly.

I heard a rustling in the elephant grass in front of us when a 400-pound Bengal tiger leaped from the grass and began slowly walking towards us, crouched and snarling, ready to attack. He came to within twenty feet of us and snarled.

Tim, Santoosh, and I huddled together and looked as big as we could. Tim is 6'8" and 300 pounds and I am 6'1" and 200 pounds. With Santoosh, we looked very large and impressive to the tiger. The great cat roared and jumped back into the high grass.

Santoosh said it was the most dangerous encounter

he had ever had. People ask if we got a picture of the tiger and we laugh because we were so scared we actually froze. Santoosh, telling us to crowd together and not move, saved our lives. The last thing we thought of was our cameras. (Whatever the last thing we thought of might actually have been the *last* thing we thought of...)

Two

THE SECOND MOST dangerous encounter happened in Kenya, Africa, at the Amboseli National Reserve. My guide James (his Christian name, a Kikuyu tribesman), and I decided to follow a herd of elephants into a wooded oasis with many large trees and ravines.

As we approached the herd, we saw a large bull elephant standing about twenty yards from a ravine and looking at us but not reacting to our presence. The bull appeared to be in musth, which James said caused the bull to be more aggressive than usual. He was looking for a mate and would fight all who entered his domain.

The bull elephant did not show any aggression but just stood, looking in our direction and raising his trunk as if to smell the air. I asked James if we could get closer so I could get a close-up picture of this magnificent animal's eye and his artistically patterned wrinkled skin.

James said that a bull elephant in musth is extremely

dangerous, but he knew about my experiences with wildlife and my ability to keep a level head in stressful situations. He decided we could try and get closer, but we had to do it slowly. James and I had been approached by a large lioness while on foot in the bush, and he was impressed that I knew not to run and that I also stared down the big cat as she sniffed us and moved on.

James and I approached the bull elephant, walking along the ravine's edge. When we were within fifteen yards of the great beast, I raised my camera and snapped a picture, framing the eye and surrounding skin. As the camera snapped, the bull elephant raised his trunk and bugled with such force, I thought my eardrums would break.

I don't know if it was the click of the camera or the raising of my arms up to my face, but the bull must have thought we were a threat. After the bugling, he lowered his trunk, tucked it under his mouth, and began to charge. An elephant can weigh up to three tons, and an average person would think that the largest land animal on the planet would be slow. Right? Wrong. The bull charged with such speed that I thought I felt dirt kick up on the back of my neck when I turned to run.

James had already called retreat and all I saw of him was the edges of his shoes as he dove over the edge of the ravine. As I threw myself over the ravine's edge, I saw some large fallen trees lying against the side of the wall of the ravine, and I dug myself under one of the logs, trying to hide from our attacker.

My body was shaking with fear, and I could hear the elephant standing beside the logs at the top of the ravine, sniffing and bugling. I looked over to the other side of the fallen trees and there was James, cowering just like me and holding his index finger to his lips, telling me to be quiet.

Controlling my breathing was extremely difficult since I was huffing and puffing like I had just finished a marathon run, instead of a fifteen yard dash. When all grew quiet and I thought the bull elephant had gone, I felt a tremendous thud against the logs above me. With every thud that followed, dirt and the logs began to fall apart around me. The bull elephant had discovered that we were under the logs and began kicking them with his foot.

My fear returned, almost totally paralyzing me. This probably saved my life as I laid perfectly still. The bull began to bugle and kick harder, then he began ramming the few logs left over me with his trunk and head.

The attack stopped suddenly and I could now see the elephant's underbelly as he straddled the few remaining logs directly above me. Then he walked back to his herd, strutting proudly, as he had defeated another challenge to his domain.

I looked over and saw James slip out of his hiding place and slide on his rear end to the bottom of the ravine, and I soon followed him. When we reached the bottom, I saw James turn. He had a huge grin on his face. I started laughing and we hugged each other and decided to follow the small creek at the bottom of the ravine out of dangers way.

When we got back with the rest of the safari they all wanted to know why we were covered with mud and vegetation. We had quite a story to tell that night.

One

My most dangerous encounter happened about eighty miles off the coast of Mexico, in the Pacific Ocean. I was with a group of divers filming and photographing blue sharks when a large Mako shark appeared at a distance. The Mako shot through the water, picking up pieces of bait and chum from the open ocean. Mako sharks are the fastest sharks and can catch and kill sail fish and tuna. Their teeth are dagger-like and fishermen consider them the most aggressive and dynamic sport fish to catch. Mako sharks can leap out of water as high as twelve feet, and there are stories of Makos leaping into the boat of a fisherman that has hooked them, attacking the offender.

They are considered one of the most dangerous sharks in the ocean and have been known to attack divers. To film or take pictures of them, we were in a shark cage made of steel. The cage was tethered to the boat with a rope.

The cage was fifty feet away from the boat and twenty feet below the ocean surface. To get to the cage, we had to swim through the other sharks that had been baited up to the boat with chum.

We started taking pictures of blue sharks when the Mako appeared. The cage had large portholes from which to take pictures, but every time the Mako came close enough for a picture, he literally flew past the port holes of the cage so fast he was a blur.

I had to make a decision. If I wanted a picture, I would have to go outside of the cage to try to catch the shark before he accelerated. When I opened the door to the cage, I felt a hand on my shoulder. It was one of the divers in the cage giving me a wave of his index finger warning me not to chance leaving the cage.

With the regulator in my mouth, I smiled and gave him the thumbs up sign, which he reluctantly returned. As I began to exit the cage I looked at my camera to adjust the footage I thought I would need (five to seven feet) and grabbed my strobe light. The strobe was not on its bracket, but only connected to the cord because I like to angle my strobe for better lighting.

After I checked my equipment, I exited the cage, staying close to the door but far enough away to catch the Mako as he angled in for the bait. When I positioned myself, floating as if I was in space, I immediately spotted in the distance and off to my right what appeared to be a shark.

Makos and blue sharks blend in to their surroundings, camouflaging themselves against the blue of the Pacific Ocean, just like a copperhead snake blends into the leaves on the forest floor. The shark disappeared but suddenly from the same area I saw it again, a blur of movement as it approached the bait in front of the cage.

The shark moved with quick and jerky movements, which meant that it was a Mako. As it raced towards the bait, coming from my right side, I positioned myself for the best chance at getting a picture of this blue lightning bolt.

The shark came straight at me with a slightly right angle approach. At the last second, I saw that I had floated into the chum: the Mako was coming in to get a piece of chum, and I was between him and the chum.

When the Mako was about seven feet away from me, I fired off my strobe light and snapped a picture. The strobe light startled the Mako, who suddenly opened his cavernous mouth of razor sharp daggers and lunged towards my face.

I shoved the strobe light and camera into the Mako's mouth and he violently bit down on them, shaking the camera, strobe, and me. The Mako shook me back and forth like a dog playing with a chew toy. Then, suddenly, he released the equipment. I punched him in the belly with the damaged camera and strobe, and he swam away over the top of the cage.

I swam as fast as I could back to the cage and looked at my damaged equipment—upset that he had chewed up camera and strobe. When we got back on the boat, I took my equipment up to a seat at the back of the boat and began checking out what could be salvaged. I noticed my hands were shaking, and I heard the other divers talking about the Mako encounter.

I felt the hair on the back of my neck begin to raise up and I understood how close I came to having my face bitten off. I had to sit down and say a prayer of thanks for surviving. This encounter was totally my fault. I should have never fired a strobe into a large predator's face. I was a guest in someone else's home, and I did not follow the rules. I had not respected the Mako as a living creature that would react in a defensive mode to me being in his feeding area.

To this day, this incident sends chills down my spine. All these encounters happened because I placed myself into a situation that caused an animal to protect itself, its domain, or its herd. I learned from each encounter, and I hope anyone who reads these stories does too.

e • i • g • h • t

Buying And Caring For Exotic Animals: A Guide

Bringing Up Baby:
The Trouble Begins

When an average person decides to buy a pet, he or she rarely purchases anything but a dog, cat or fish. There is, however, a growing part of the population that craves something different and exotic—a pet that will draw attention to the owner, one that is different from the normal pet. There are even people who collect venomous animals, as we have learned, and keep them in their homes. I would like to close my book with some helpful advice on how to choose a pet that will be perfect for what you are looking for, be it exotic or domestic.

First, go to a library or bookstore and find a book on the animal that has caught your interest. When television shows depict exotic animals as adorable, the shows usually neglect to mention how difficult it is to train an animal to be that way—or how many stitches the trainer has had.

Pick up a book on your selected animal and learn how the animal lives in the wild, where it is supposed to be in the first place. The tiger is a popular pet in America. Most good-willed people think they simply get a cub and raise it just like on television, playing with it and letting it live in their homes.

But ninety percent of the big cats are either shot by law enforcement officers when they escape, or they are euthanized because no zoo or nature park wants them.

Most big cats have been de-clawed and are in poor health because of the owner's lack of knowledge about nutrition. A de-clawed, neutered tiger cannot be enclosed with a healthy, fully intact tiger because the defenseless cat will most likely be killed.

Then there is the owner who starts to raise the cub and when "baby" reaches a hundred pounds, they learn it no longer behaves like the cuddly tiger on the animal television show. The owner gets bitten and his or her home is wrecked. The worse scenario is that the tiger seriously hurts or kills a friend, neighbor, or family member and then the cat has to be killed.

After that, the real trouble begins. Attorneys get involved, suing the owners of exotic pets because of their animal's mistakes. "You studied all about your exotic pet before you brought such a potentially dangerous animal into your suburban neighborhood—where children play free of fear of being attacked by a tiger—didn't you?" the attorney begins. Many an exotic pet owner has lost everything—home, car, family, and income—over the desire to have a unique pet.

I have heard may stories of wolves, big cats or bears turning on their owners. The words "turning on their owners" is used often in newspaper articles or television newscasts. I have learned from my years of experience that these animals do not "turn on" the owners. They react as they are supposed to do if they were still in the wild, to a situation that the owner has presented unknowingly.

An example. A wolf owner is pinned to the ground by his 160-pound male wolf. The owner called me, crying that he would have to put his wolf to sleep because of the aggressive behavior, because it "turned on" him. I talked to the owner and found he had just recently injured his knee and was on crutches, hobbling along and in pain. When he came home from the doctor and entered his house, his wolf jumped him, grabbing him by the throat and pinning him to the floor, snarling and growling.

The owner acted correctly—he laid perfectly still and did not try and fight back. The male then let go and continued as if nothing had happened. I explained to the owner that what had happened was not that the wolf "turned on" him but that the wolf, a little over two years old, had reached sexual maturity and considered the owner the Alpha male.

Wolves, big cats, and bears all have a hierarchy in their lives in the wild and knowledge of this can save an owner's life. It can also prevent the needless deaths of animals who are merely doing what they were created to do. All react aggressively and this aggression can be triggered by many different stimulations. The most notable, appearing in over ninety percent of all attacks, is the use of alcoholic beverages. The smell triggers an interest by the animals and alcohol also causes the subject's body language to change drastically. This makes the animals become aggressive, provoking them to test the owner.

Read about your potential pet. Study about how it matures, what its final size will be, what it eats, and how it kills its prey. If you have a venomous animal, what kind of venom does it have—neurotoxic or hemotoxic or a mix of both? What hospital in your area knows the lifesaving procedures needed to save you if you are bitten by your venomous pet? And, what do you do with your exotic pet when it does not live up to your expectations. Please note: No one ever gets to ask the animal about *its* expectations. Animals live in a distinctly homocentric world.

When purchasing an exotic pet:

1—Read about your proposed pet; where they live in the wild, potential size and disposition, what and how much they eat, dangers the animal might pose to you and your other pets.

2—Call your local zoo, nature center, or museum. If you're still in school, check with your teacher. These professionals can help you understand about your interest and they might let you volunteer at their facility.

3—Talk to people that have raised or kept the animal in which you are interested, and ask how to cage the animal, what to expect from neighbors, and what kind of problems they experienced with their first exotic pet.

4—If you can, correspond with a researcher from the country your proposed pet originated from, such as a tiger. Research from the Tiger Project in India can teach you first hand about these animals. This information will usually convince you *not* to have a tiger in your home but to leave them in the wild. If a person really loves tigers or exotic animals, this decision will be easy to make.

In today's computer age, information is easy to find. If you live in a small apartment or a suburban area, making a decision about the kind of exotic pet to have is similar to purchasing a dog. A Great Dane will not fit comfortably in a one bedroom apartment. Neither will a six foot alligator or a two hundred pound black bear.

It helps to understand that an adorable two foot snake will grow into a mammoth twenty feet and weigh over 150 pounds. A snake this size is no longer a pet, it is an ordeal. If you are interested in owning a snake, start with a locally-known snake. Black rat snakes or corn snakes make great pets. If an owner gets tired of them, he can turn them loose and not worry about them dying. These snakes rarely bite, make low maintenance pets, and usually do very well in captivity.

As your experience and knowledge grows, you can start looking into small growing boas and pythons. A Ball python or boa constrictor can make an interesting exotic pet and grow an average of six to ten feet. They need heated and specially lighted cages but are usually easy snakes to maintain.

I advise novices to stay away from anacondas, Burmese pythons, reticulated pythons, and any other snake that can reach a length of twenty feet and weigh close to two hundred pounds. These snakes are not usually dangerous, but there have been occasional deaths from each of these snakes. Their size makes it difficult to have the proper cages and to move such a large animal. Feeding is expensive and it is nearly impossible to get rid of a snake that is over fifteen feet. This causes many owners to use drastic measures. They either turn it loose or kill it.

A novice should not own venomous snakes. Even professional snake-keepers have had near death experiences after being bitten by their venomous snakes. Only professionals with years of experience and actively using the snakes for programs or venom research should own venomous snakes. If the venomous snake owner does not have access to antivenin serum, he should not be around such a dangerous animal.

I have had many people complain that I should not preach about regulations on venomous or exotic animals. They say they should have the right to own a tiger or cobra. The unfortunate problem is that the tiger and the cobra also have rights, and the law as it is today affords these creatures little or no protection. In turn, they may endanger their keepers, as well as others.

If people can't use common sense on their own—which is usually the case, to wit, the incident of the spaghetti-loving bear in the apartment complex—then regulations need to be in effect to protect those concerned. Much of the time, those who need protection most are the animals, who are endangered by their keepers.

I have met many people who have built beautiful facilities for bears and big cats, and their animals are happy, clean, and healthy. Unfortunately, these people are rare. Most of these owners have the animals living in their small homes, or in miserable cages, unhealthy and aggressive.

Before a person decides to purchase a big cat or bear, they should have a facility that is large and easy to clean and maintain. They should have a veterinarian that will help them with the exotic pet's health needs. The person needs to understand local regulations for having such animals, and to talk with neighbors. It is better to have them on your side than to have them opposing you.

When you have covered all these angles, you are ready to decide what kind of animal you should get. Tigers and lions roar and growl, and sometimes can be heard as far as seven miles away. Bears smell awful. If you still want a big cat or bear, finding a reputable dealer can also be difficult. There are many animal dealers but few good ones.

Personally, I am against people having big cats and bears in their homes. These magnificent animals need to be in the wild. I have raised big cats and bears but after traveling to the countries where they originated and seeing them in the wild, I have come to regard captivity as a sadness for them. Caging a tiger is cruel. Forcing a tiger to do tricks in a circus strips the dignity and the respect of this regal animal. If a prospective owner saw these animals first in their natural element, one might think twice about ownership.

There are many lizards that an exotic animal owner could have and enjoy without much expense or danger. Bearded dragons, geckos, chameleons, anoles, and smaller lizards make excellent pets, but the prospective owner needs to gain as much information as possible before purchasing one.

Other lizards are more problematic. Be wary of monitor lizards because they can grow to as much as nine feet and be aggressive and dangerous. Be wary of venomous lizards such as Gila monsters and beaded lizards, neither of which make good pets. Large iguanas may bite and they carry salmonella.

The prospective pet owner needs to understand that some animals never make good pets, that others may require special lighting and heating which can also drain your pocketbook, and some species, such as tortoises and parrots can outlive their owners—which may be a problem for family members who are not interested in inheriting an ancient turtle or parrot.

More people are buying alligators, caimans, and crocodiles as novelties, and when they get big enough to bite, the owners lose interest and get rid of them. Alligators can quickly grow up to ten feet long. Crocodiles can attain a length of fifteen feet. These are not pets. These are dangerous animals.

I do not advise anyone reading this book to purchase a crocodile. If you wish to work with these critters, contact your local zoo to volunteer or travel to where these animals are living naturally and volunteer with researchers. Enjoy these animals where they were meant to be, in the wild.

I hope my Do's and Don'ts of exotic pets will help the reader to make an educated decision on what kind of animal makes a good pet. I also hope my reservations may even keep people from buying any exotic animal. With more than thirty years of experience raising and rehabilitating wild animals, I have met only a few people who have had good experiences keeping exotic animals in their homes.

Many of the stories in this book have been sad stories of humans being hurt and animals being killed. The television shows are filled with people using animals circus-style, yanking them by their tails, showing off with venomous snakes, and disrespecting the animals, all because of the host's ego trip.

These ego trips have caused reverse problems with animal lovers. We went through a time when people respected wild animals and wanted them left alone in their element. Now irresponsible television shows depict hosts pulling animals from trees and holding them by their tails. They show snakes and alligators biting cameras and damaging the delicate facial organs in their faces and mouths. The hosts jump on the backs of an obviously sedated or overfed crocodile and wrestle it into submission. These animals are purposely teased to hiss and attack the camera, providing cheap and manipulated thrills for a gullible audience that thinks the ego-stricken television personality has barely escaped with his life.

The brainless popularity of this kind of animal programming has instigated people into taking unnecessary risks while showing off for friends and family.

Statistics have shown that venomous snake bites have increased ninety percent in North America in the last decade. There are more calls for antivenin than ever before. Seventeen children are bitten by venomous snakes each year. Ten years ago there were only two.

Most of these bites are from exotic pet snakes, whereas in the past, most snake bites came from local snakes. Male juveniles have also received serious injuries from alligators. One boy from Louisiana lost an arm trying to imitate his favorite wildlife television host.

I hope this book will help people to understand that wild animals do not want to be in your home. If you love exotic animals, you will leave them in the wild. Believe only a small percentage of what you see on television. Go out and experience the rest for yourself.

Appendix

Snakebite Protocols, Or: What To Do Until The Priest Arrives

My brother Jim and I have often helped doctors and medical personnel when they have had to treat a venomous snakebite victim. Jim is an internationally known venom extractor and has given lectures all over the world on snakebite treatment. He and Dr. Sherman Minton have developed a snakebite protocol to follow and assist medical experts in successfully treating the victim. We do not treat patients. We consult with doctors to identify the snakes that have bitten the patient. A small story will illustrate the importance of a good snakebite protocol. A few years ago, a man was bitten by a snake and rushed to the hospital nearest him. The family cut off the snake's head and brought it along, in a shoebox. The attending physician looked at the snake, which had been damaged in the assault, then got out his book of snake drawings. The snake, he was certain, was a copperhead. He put a tourniquet on the man's arm, which cut off the circulation, causing it to swell. To reduce the swelling, the doctor stuck the man's arm in a bucket of ice, and it was kept there for an inordinately long time. This, of course, caused the arm to look worse, so the physician did a fasciotomy, cutting along the extremity to reduce the swelling. In the process, the arm became infected, antibiotics were administered, to no avail, and the man lost his arm. The snake was later identified as a garter snake. Only a few years ago, a tourniquet was considered proper treatment and ice was used for swelling, both considered old wives's tales in modern treatment. But, still, all snakebites are not considered venomous. Physicians, themselves, sometimes do not understand that a majority of even venomous bites are "dry" bites, in which the snake injects little or no venom. This necessitates observing the victim closely and treating the symptoms. The following is what we consider a successful protocol.

Venomous Snakebite Protocol

I. Secure snake.

~ Secure snake in locking container or room. It is crucial to know what kind of snake has bitten you.

~ Euthanize if too difficult or dangerous to secure.

~ Post guard and call for help if the above is not possible.

II. Identify the type of bite

A. Elapid, neurotoxic pit vipers, viper bites

~ king cobra, *Ophiophagus*

~ cobra, *Naja*

~ shield-nosed cobra, *Aspidelaps*

~ krait, *Bungarus*

~ mamba, *Dendroaspis*

~ Mojave Central, South American rattlesnake*, *Crotalus durissus*, *Crotalus horridus* Type A

*Pit vipers or viper species possessing highly eurotoxic venom

B. Viper or pit-viper bites

- Gaboon, *Bitis*
- rattlesnake *Crotalus*, excluding *durissus*, *horridus* Type A
- fer-de-lance, South American pit vipers, *Bothrops*
- sawscaled viper, *Cerastes*, *Echis*
- cottonmouth, copperhead, cantil, *Agkistrodon*
- Russell's vipers, *Daboia*, all *Vipera* without highly neurotoxic venom

III. First-aid

A. Elapid, neurotoxic pit vipers, viper bites

1. Obtain pressure bandage from first-aid kit, start at bite and wrap tightly up entire limb (i.e. start at hand and wrap to shoulder)
2. Immobilize limb (splint wrapped to limb with bandage)
3. Artificial ventilation (tracheotomy) may be required
4. Transport to hospital immediately

B. Viper or pit viper bites:

1. No first aid; transport to hospital immediately
2. Watch for, treat anaphylaxis (allergic reactions)

IV. Transport to hospital

- Ideally one person should drive while another monitors/treats patient
- Be sure to take antisera and anaphylaxis kit

A. Elapid bite:

1. Transport to any hospital

2. Notify hospital (while patient is en route, if possible)
3. Instruct hospital to prepare helicopter (to obtain antivenin)

☎ *Help Numbers for Hospital Personnel:*

Toxin office: 513-281-8094

Local EMS
Dr. Barry Gold: 410-484-5640 (w)
Dr. Otten: 513-281-8574 (w)

Miami-Dade Fire Rescue, Florida
Capt. Al Cruz, Florida Antivenin Bank: 786-331-5000
http://www.co.miami-dade.fl.us/firerescue

Jim Harrison, Kentucky Reptile Zoo and Captive Born Venom Lab: 606-663-9160
Tim Harrison, City of Oakwood, Ohio, public safety: 513-298-2122

B. Viper bite:
1. Transport to nearest hospital
2. Notify hospital (while patient is en route, if possible)

V. Contact support network, for acquiring antisera

A. Additional antisera:

☎ *Dr. Richard Dart: 303-629-1123*
(Colorado Poison Control)

Johnny Arnette: 513-281-4701, ext. 8355
(Cincinnati Zoo)

Jim Harrison, Kentucky Reptile Zoo and Captive Born Venom Lab: 606-663-9160

Tim Harrison, City of Oakwood, Ohio, public safety: 513-298-2122

B. Surrender the antisera only to the administering physician

C. Assistance in the monitoring of treatment, and securing snake:

☎ *Tim Harrison, City of Oakwood, Ohio, public safety: 513-298-2122*

Jim Harrison, Kentucky Reptile Zoo and Captive Born Venom Lab: 606-663-9160

Or local animal control

Hospital Treatment

~ No ice

~ No tourniquets

~ No cut and suck

~ No Cortisone

~ No surgery (no fasciotomy or debridement)

I. Start IV

II. Watch for these systemic effects and do complete blood workup:
- ~ nausea or vomiting
- ~ abnormal blood pressure
- ~ involuntary muscle reactions (face and hands)
- ~ drooping eyelids
- ~ seizures
- ~ dementia
- ~ unconsciousness or coma
- ~ dementia
- ~ shock
- ~ any other abnormalities

III. Administer antisera if needed (indicated by presence of symptoms)
- ~ Administer intravenously only
- ~ No trial dose necessary (administer first vials via slow drip IV)
- ~ Administer 5-10 vials, increase dosage as needed

IV. Be prepared to treat for anaphylaxis
- ~ After antisera administered, watch for anaphylactic reaction
- ~ Treat with adrenaline

V. Administer antibiotics after antiserum has neutralized venom
- ~ IV: Keflix

- Oral: Keflin, Cipro, or Augmentin. Culture bite site within 24 hours.
- Give plasma if needed following venom neutralization

VI. Observe for 72 hours

Epilogue

THIS BOOK WAS WRITTEN TO INFORM the general public of the increasing chance that they will be coming across an exotic or local wild animal as they increase in many suburban areas across the county. New homes built in wooded areas cause city folk to run into local critters they have never seen or knew existed, and they respond in panic to their local police department where the majority of police officers or even animal control officers have never dealt with such creatures, either.

My favorite occasion is responding to a neighborhood where a large buck deer and his herd are in someone's yard, and the plat of houses in which these people reside is called Deer Run Estates. Truth, inadvertently, in advertising. There was a large snapping turtle in the front yard of a new residence, and I was called to capture the turtle and relocate it. When I arrived, I saw that the name of the plat of new homes was Turtle Crossing. How could I possibly keep a straight face when the property owner kept saying, "I can't believe a predator like that exists in Ohio."

Perhaps we should put more realistic place names on new subdivisions: Lawyers Run, for instance, or Accountants Crossing. I don't think people are stupid, they forget that they, too, live in niches. And if they move into a new area, they should go to a library or the local nature center and learn about their animal neighbors.

These same people have checked the schools, police and fire coverage, and zoning codes. Why not find out if their property is in the migration path of deer, or if their home is built on a marsh area where animals have been coming for food and cover for hundreds of years.

There is a story where a couple moved to Kentucky and built a beautiful home in a wild area. The husband hated snakes and when he saw an eastern king snake in his yard, he wanted it moved immediately. A local reptile expert came to the property and tried to talk the owner out of moving the king snakes off his property. The expert told him that the king snake was important. If it were removed, the owner would have broken a link in the chain of life, causing the natural balance of this property to fall apart. The owner ordered the all snakes removed from his property immediately.

The snake expert did what the owner wished and removed many snakes, mostly eastern king snakes. Three months later, the homeowner frantically called the snake expert. He had now discovered a copperhead snake near his home.

The snake expert told the homeowner that removing the king snake caused a negative situation; the king snake eats copperheads and rattlesnakes and, as he had explained before, the removal of the king snakes might cause venomous snakes to prosper.

When the snake expert arrived back at the residence, he brought with him some king snakes to release on the property. First, the expert collected copperheads and a large timber rattlesnake near the home. Then he released the king snakes back into the yard area, and the chain of life was complete again. The venomous snakes were taken to a secluded area and released. There were no more calls of snakes in the yard at that home again.

The owner of that property was now educated, and although he had learned in a more difficult manner about his new environment, he had learned.

This story sums up why I have written this book. I hope the reader has been entertained and, like the Kentucky homeowner, educated painlessly.

About The Author

TIM HARRISON has been a public safety officer, performing as a police officer, firefighter, and EMT-paramedic for twenty years for the City of Oakwood in Montgomery County, Ohio. He has been involved with wildlife rescues, rehabilitation, and raising animals since he was 13 years old, and he continues to learn from every incident he is called to help on. He has traveled around the world with authors and television production crews that were either chasing unusual exotic animals or filming wildlife shows, and he has his own production team that has produced such local television shows as *The Wolf Recovery Program of Yellowstone*, *Blue Sharks of the Pacific*, *Bears of Yellowstone*, and *Great White Sharks of Shark Alley, South Africa*. In his own terms, he describes himself as a wildlife enthusiast and his philosophy is: Do no harm. He lives in Springboro, Ohio, with his wife, Patricia, and their three children, Adam, Alex, and Aric. The author has also assisted his brother, Jim Harrison, world-renowned herpetologist, in the development of the Kentucky Reptile Zoo and Venom Laboratory in Natural Bridge, Kentucky.